# SECURITY IN SUSTAINABLE ENERGY TRANSITIONS

By providing a new qualitative analysis of policy coherence and integration between energy, security, and defence policies between 2006 and 2023, this book analyzes the impacts of policy interplay on energy transition through the lens of sustainability transitions research, security studies, energy security and geopolitics, and policy studies. The security aspects discussed range from national defence and geopolitics, to questions of energy security, positive security, and just transitions. Findings show that the policy interface around the energy–security nexus has often been incoherent. There is a lack of integration between security aspects, leading to ineffective policies from the perspectives of decarbonization and national security, evident in the European energy crisis following the war between Russia and Ukraine. This book is intended for researchers and experts interested in the energy transition and its connections to security and defence policies. This title is also available as Open Access on Cambridge Core.

PAULA KIVIMAA is Research Professor in Climate and Society at the Finnish Environment Institute and Associate in the Science Policy Research Unit at the University of Sussex. She is an internationally renowned scholar in sustainability transitions, a coeditor of the book *Innovating Climate Governance* (Cambridge University Press, 2018), and a member of scientific committees advising the European Commission and the Finnish government.

# SECURITY IN SUSTAINABLE ENERGY TRANSITIONS

Interplay between Energy, Security, and Defence Policies
in Estonia, Finland, Norway, and Scotland

PAULA KIVIMAA
*Finnish Environment Institute*

CAMBRIDGE
UNIVERSITY PRESS

# CAMBRIDGE UNIVERSITY PRESS

Shaftesbury Road, Cambridge CB2 8EA, United Kingdom

One Liberty Plaza, 20th Floor, New York, NY 10006, USA

477 Williamstown Road, Port Melbourne, VIC 3207, Australia

314–321, 3rd Floor, Plot 3, Splendor Forum, Jasola District Centre, New Delhi – 110025, India

103 Penang Road, #05–06/07, Visioncrest Commercial, Singapore 238467

Cambridge University Press is part of Cambridge University Press & Assessment, a department of the University of Cambridge.

We share the University's mission to contribute to society through the pursuit of education, learning and research at the highest international levels of excellence.

www.cambridge.org
Information on this title: www.cambridge.org/9781009368162

DOI: 10.1017/9781009368155

© Paula Kivimaa 2024

This publication is in copyright. Subject to statutory exception and to the provisions of relevant collective licensing agreements, with the exception of the Creative Commons version the link for which is provided below, no reproduction of any part may take place without the written permission of Cambridge University Press & Assessment.

An online version of this work is published at doi.org/10.1017/9781009368155 under a Creative Commons Open Access license CC-BY-NC-ND 4.0 which permits re-use, distribution and reproduction in any medium for non-commercial purposes providing appropriate credit to the original work is given. You may not distribute derivative works without permission. To view a copy of this license, visit https://creativecommons.org/licenses/by-nc-nd/4.0

When citing this work, please include a reference to the DOI 10.1017/9781009368155

First published 2024

A catalogue record for this publication is available from the British Library

Library of Congress Cataloging-in-Publication Data
Names: Kivimaa, Paula, author.
Title: Security in sustainable energy transitions : interplay between energy, security, and defence policies in Estonia, Finland, Norway, and Scotland / Dr Paula Kivimaa, Finnish Environment Institute.
Description: Cambridge, United Kingdom ; New York, NY : Cambridge University Press, 2024. | Includes bibliographical references and index.
Identifiers: LCCN 2024011636 | ISBN 9781009368162 (hardback) | ISBN 9781009368155 (ebook)
Subjects: LCSH: Energy industries – Law and legislation – Europe. | Renewable energy sources – Law and legislation – Europe. | Carbon dioxide mitigation – Law and legislation – Europe. | National security – Law and legislation – Europe. | Military readiness – Law and legislation – Europe.
Classification: LCC KJC6848 .K58 2024 | DDC 333.79094–dc23/eng/20240317
LC record available at https://lccn.loc.gov/2024011636

ISBN 978-1-009-36816-2 Hardback

Cambridge University Press & Assessment has no responsibility for the persistence or accuracy of URLs for external or third-party internet websites referred to in this publication and does not guarantee that any content on such websites is, or will remain, accurate or appropriate.

# Contents

*Preface* *page* ix
*Acknowledgments* xi

1 Introduction: The Challenge of Zero-Carbon Energy Transitions
  and National Security 1
  1.1 Conceptual Background 5
  1.2 A Small Country Perspective: Estonia, Finland, Norway,
      and Scotland as the Research Foci 8
  1.3 Research Method and Materials 10
  1.4 Contents of the Book 12

**Part I Theoretical and Literature-Based Foundations**

2 Understanding Security in the Context of Sustainability Transitions 17
  2.1 Sustainability Transitions Research: Key Conceptualizations 18
  2.2 Conceptualizing the Basics of Security for Sustainability
      Transitions 25
  2.3 Security in Transitions Research 31

3 Energy Security and Geopolitics of Energy Transition 35
  3.1 Conceptualization and History of Energy Security Research 35
  3.2 Geopolitics of Renewables 39
  3.3 Energy Security in Europe 48

4 A Conceptual–Analytical Approach to Examining Security
  in Sustainability Transitions and Policy Interplay 52
  4.1 Security as Part of the Sociotechnical Landscape
      for an Energy Regime 54

|  |  |  |
|---|---|---|
| | 4.2 Policy Coherence at the Regime Level: Interplay of Energy Transition Policies with National Security and Defence Policies | 58 |
| | 4.3 Security in Change Processes: Niche Expansion and Regime Decline | 63 |

**Part II  Empirical Case Studies**

| | | |
|---|---|---|
| 5 | Estonia: Long-Term Energy Independence and Oil Shale | 69 |
| | 5.1 Energy Regime | 72 |
| | 5.2 Security Regime | 76 |
| | 5.3 Perceptions of Russia as a Landscape Pressure at the Intersection of Energy and Security | 79 |
| | 5.4 Policy Coherence and Interplay | 80 |
| | 5.5 Niche Development, Regime (De)stabilization, and Positive and Negative Security | 84 |
| | 5.6 Concluding Remarks | 92 |
| 6 | Finland: Ambivalent Links between Energy and Security | 94 |
| | 6.1 Energy Regime | 96 |
| | 6.2 Security Regime | 102 |
| | 6.3 Perceptions of Russia as a Landscape Pressure at the Intersection of Energy and Security | 103 |
| | 6.4 Policy Coherence and Interplay | 107 |
| | 6.5 Niche Development, Regime (De)stabilization, and Positive and Negative Security | 111 |
| | 6.6 Concluding Remarks | 117 |
| 7 | Norway: Contradiction of Oil for Export and Fully Renewable Electricity Supply | 119 |
| | 7.1 Energy Regime | 122 |
| | 7.2 Security Regime | 127 |
| | 7.3 Perceptions of Russia as a Landscape Pressure at the Intersection of Energy and Security | 128 |
| | 7.4 Policy Coherence and Interplay | 129 |
| | 7.5 Niche Development, Regime Stabilization, and Positive and Negative Security | 131 |
| | 7.6 Concluding Remarks | 138 |
| 8 | Scotland: From Oil to Wind under a Devolved Government and New Pressures for UK Energy Security | 141 |
| | 8.1 Energy Regime | 144 |

|  |  |  |
|---|---|---|
| 8.2 | Security Regime | 149 |
| 8.3 | Perceptions of Russia as a Landscape Pressure at the Intersection of Energy and Security | 151 |
| 8.4 | Policy Coherence and Interplay | 153 |
| 8.5 | Niche Development, Regime Destabilization, and Positive and Negative Security | 158 |
| 8.6 | Concluding Remarks | 160 |

**Part III Conclusions**

9 Insights into Zero-Carbon Energy, Sustainability Transitions, and Security  165
   9.1 Interplay between Energy, Security, and Defence Policies  165
   9.2 Securitization and Politicization of Energy Transitions  169
   9.3 Security Implications of Energy Transitions  170
   9.4 Technological, Actor-Based, and Institutional Aspects  174
   9.5 Further Insights for Sustainability Transition Studies  181
   9.6 Final Remarks  183

*References*  187
*Index*  211

# Preface

Until the turn of the 2020s, questions around security and defence were not part of sustainability transitions research and were also little addressed in the context of energy transitions. The first two decades of the research field of transitions had an innovation bias. It was built around the search for knowledge on how radical innovations emerge and gradually transform sociotechnical regimes around energy, mobility, or food. Only since the 2010s has more transitions research begun to address the acceleration of transitions and how the processes of scaling innovations connect to the decline of sociotechnical regimes. The established sociotechnical systems have often been described as suffering from path dependence and lock-in, and, hence, destabilization or decline of such systems is difficult to realize. Alongside the attention to regime decline, discussion on broader repercussions of transitions, such as social justice, emerged. Yet little attention was paid to security and defence as influencing factors on transitions or in terms of transitions shaping the security landscape.

My own motivation to research the topic of this book emerged when I heard Professor Johan Schot talk in 2017 about how the sustainability transitions community has paid scarce consideration to negative side effects and the role of war in transitions. This influenced me, as a policy-oriented transitions scholar, to think about defence policy and its interconnections with sustainability transitions. As a result, I wrote a research proposal on this topic, with an empirical focus on the zero-carbon energy transition and its security and defence connections. One of the key areas of interest was policy coherence and integration between energy and defence policies from a transitions perspective. The proposal was awarded funding from the Research Council of Finland in 2019.

Soon after I began the project in September 2019, I realized that a focus solely on energy and defence is too narrow and that a broader perspective on security and security policies is likely to be more fruitful, going also beyond energy security. Therefore, this book takes a broad perspective on security and how it connects to sustainability transitions. It departs from more traditional national security perspectives tied into defence and geopolitics. It also addresses energy security via security of supply, energy independence, and the operational security of electricity grids. Finally, it touches upon other dimensions of security, such as climate security, positive security intertwined with questions of internal stability of countries and of just transitions, and cybersecurity. The book is based on an analysis of almost ninety expert interviews and over a hundred policy documents. Instead of predefining security, this was left open to be defined by the interviewees and the framing of the policy documents. Nevertheless, for the reader, the book also offers conceptualizations presented in security studies and in energy security research as signposts to the diversity of ways in which security can be understood.

During the initial stages of research, it turned out that few had thought or knew anything about the interconnections between energy transitions and security. In each case country, only about a dozen people could be identified in this interface, most also not considering themselves as experts. The general awareness about energy and security has increased following the high media and policy attention given to the topic after Russia attacked Ukraine in 2022 and an energy crisis in Europe resulted. This book makes detailed knowledge on the energy transition–security nexus more widely available.

This book is intended for researchers and experts interested in the energy transition and its connections to security and defence questions. It is also addressed to sustainability transition researchers in other domains. The book combines sustainability transitions research, security studies, energy security and geopolitics, and policy studies in order to ground qualitative empirical analysis. It focuses on energy transitions and security, and the interplay of energy, security, and defence policies in four European countries in the period 2006–2023, starting from the year of the first Russia–Ukraine gas crisis. Due to the timing of the analysis, the book is able to document in detail the interconnections before and after the 2022 shift in the European energy landscape. It, therefore, also offers insights to decision-makers interested in the past and present of energy transitions and security policies.

# Acknowledgments

This book has been written as part of the project "IDEALE Interplay between National Defence, Security and Low-Carbon Energy Policies: A Sustainability Transitions Perspective." It has been funded by the Research Council of Finland (decision numbers 322667, 328717, and 352972) and the Finnish Environment Institute Syke.

I am grateful to many colleagues for their contributions to different parts of the project, making this book possible. First, I want to thank Johan Schot, in the capacity of my former foreperson in the Science Policy Research Unit (SPRU), for influencing the formation of the project idea and supporting the development of the project proposal. My thinking on the potential negative side of sustainability transitions and the impact of defence policy on energy transitions was inspired by Johan. Second, I want to thank Marja Helena Sivonen, for being a lovely project partner and friend with whom to explore the topic of this book. She has gathered the empirical material with me, commented on the contents of the book, and been the coauthor in several peer-reviewed scientific articles that the project has produced. Thank you also to Claire Mosoni, who, as a part-time project member, provided information for the Estonia chapter and read through and commented on the other chapters.

I want to thank several people for commenting on different versions of this book's chapters. I am indebted to Suvi Huttunen for reading the whole book draft at the end and providing excellent comments. My thanks go also to Matthew Hannon, Phil Johnstone, Laur Kanger, Jukka Leskelä, Jani Lukkarinen, Marianne Ryghaug, Ben Shafran, Gert Siniloo, and Veli-Pekka Tynkkynen for providing valuable comments on selected chapters. I, of course, remain responsible for any remaining inaccuracies. I also want to thank Bipashyee Ghosh, who helped to create an index for the book, and Satu Turtiainen, for making the figures look professional. Huge thanks go to all the people who agreed to be interviewed for this project. Without their input, the analyses carried out would not have been possible.

During the project I had the pleasure of coauthoring an article with Marie Claire Brisbois, Emma Hakala, Dhanasree Jayaram, and Marco Siddi. This process was instrumental in me getting more acquainted with security studies and becoming aware of the concept of positive security, on which I draw a great deal in this book. I have also had many discussions with colleagues in the European Union Cascades project and the European Consortium for Political Research Energy Group that have informed my thinking. I want to thank my former and current forepersons, Mikael Hildén and Johan Munck af Rosenschöld, for their valuable support, and my family for the joy and love they have given during the writing of this book. Finally, I want to remember a dear colleague, David Lazarevic, who passed away on September 1, 2023. He was my go-to person in the Finnish Environment Institute to debate and deliberate concepts and issues related to sustainability transitions. This book is dedicated to the memory of David.

# 1

# Introduction

*The Challenge of Zero-Carbon Energy Transitions and National Security*

Energy transitions based on decarbonized energy sources, electrification, and digitalization have, since 2015, accelerated in many places throughout the world. Shares of renewable energy have increased, new climate targets and policies have been set, and new interconnections have enabled increasing electrification and cross-country electricity markets. More modestly, developments in the energy efficiency of housing and transport have advanced, and citizen energy ownership has gradually grown. The events of 2022, when Russia began a war in Ukraine, changed the energy landscape of Europe substantially. Due to Europe's dependence on Russian energy sources, the war has also meant an acceleration of the ongoing energy transition away from fossil fuels and toward the improvement of energy efficiency. This book is based on research that began in 2019 and was completed in 2023. It examines the energy transition and its security and defence connections in selected European countries before and after this changed energy landscape.

Energy transition can mean many things. This book takes the lens of electrification and expansion of renewable energy sources as its particular foci. Electrification means increasing the use of electricity and the number of its applications across different sectors. For example, transport, heating, and industry are shifting from other power sources to electricity. Electricity can be used in transport via electric vehicles, in heating directly or via heat pumps, or to produce electric fuels (e.g., e-kerosene, e-methane), via so-called power-to-x processes connecting to the hydrogen economy and used for a variety of needs. Renewable energy sources – in particular wind and solar – can be used in increasing capacities to produce electricity. For instance, globally renewable sources are expected to cover 40 percent of electricity production in 2027 (IEA, 2023). Thus, electricity will have a significant role in decarbonizing the use of energy and improving energy efficiency. This transition will require the construction of new electricity distribution and transmission capacity within and across countries.

There are expectations of regional grid communities that will arise between countries with shared electricity networks and markets. The Nordic Electricity Market NordPool is a good example of such a community. It was initiated in 1996 with the establishment of a power exchange between Norway and Sweden; Finland joined two years and Denmark four years after. In the twenty-first century, NordPool has expanded by opening bidding areas in Germany and the Baltic countries. In 2023, in the NordPool power exchange, the share of wind power at times reached around 30 percent on the spot market, so the decarbonization process is underway but is still far from complete.

At the same time, however, the share of hydrocarbons in the world's energy production remains significant, in total over 2 million megawatts (MW). New facilities are being planned or are under construction in China, India, and elsewhere in Asia, as well as in Poland.[1] The International Energy Agency (IEA) has estimated that coal-fired power generation increased by 3 percent in 2022. Globally, the demand for energy is rapidly increasing. The IEA has forecast that electricity demand will grow more rapidly than the installed capacity of renewable energy. Further, the growing competition for energy is estimated to lead to more unstable global energy markets, with the rise of renewable energy and digital technologies driving an increasing demand for raw minerals and metals, some of which have been labelled "critical" and others also "rare earths." It has been estimated that the shift from hydrocarbons to renewable energy will substantially increase the demand for critical materials, a demand that is not wholly solvable by recycling these materials (Michaux, 2021). Thus, with the global energy transition unfolding, we see a shift from hydrocarbon-based fuel dependency to broader mineral dependency. This not only impacts global trade and geopolitics, but, via pressures to increase mining and refining, has also created new challenges for social justice, conflict reduction, and innovation – locally, regionally, and globally.

The zero-carbon energy transition based on renewables, combined with electrification, has an array of implications for different aspects of security, which is the topic of this book. These range from traditional military security concerns and geopolitics to security of energy supply, and to internal domestic security, touching upon human security, cybersecurity, and climate security. This book explores some of these security aspects from the perspectives of selected small Northern countries in Europe – Finland, Norway, and Estonia – and via the ambitions of the nation of Scotland in the UK, which has often looked to the example of the Nordic countries.

Such security implications were little discussed in Europe before 2022. Energy was mainly seen from a free market perspective, with an emphasis on open energy

---

[1] www.carbonbrief.org/mapped-worlds-coal-power-plants/ (accessed September 6, 2021).

markets as creating the most inexpensive and efficient European energy system. This stance remained until 2022, despite openings in academic research that raised concerns about the security risks of European energy systems (e.g., Scholten, 2018; Tynkkynen, 2021). Russia's aggressive military attack on Ukraine, however, brought the geopolitical aspects of energy, as well as energy and resource security, to the top of the European Union's political agenda. Suddenly, wider audiences were fully aware of the connections of energy policy and energy transition to European security.

While an alternative or a complementary pathway to the zero-carbon energy transition is presented by expanding the use of nuclear power, this book focuses mainly on the diverse security implications of renewable energy-based transitions. While the nuclear path is also relevant from the security perspective, this has been addressed elsewhere.[2] This book is concerned with the vision of future energy systems that are based on renewable energy and electrification. Such a vision has been driven, for example, by the International Renewable Energy Agency (IRENA). However, I also make some remarks regarding nuclear power as a security-related question in Finland and Scotland.

Two of the countries examined in this book, Finland and Scotland, use nuclear power as part of their energy portfolio. Finland's newest nuclear reactor, Olkiluoto 3, began operating in 2022, twelve years behind its initial schedule. It, however, suffered from several technical problems, with further delays until 2023. Finland has no direct military security interests linked to nuclear power; although Rosatom – a Russian government-owned company – was involved in Finland's new Hanhikivi nuclear power plant development, the link was an indirect one. Scotland, in turn, opposes the construction of new nuclear power plants, despite nuclear power being connected to military interests in the UK (Johnstone and Stirling, 2020). Estonia and Norway do not have any nuclear power plants. In 2011, the Estonian government approved plans for a nuclear power plant to be constructed by 2023 but this was not realized. However, a government-level working group was established in 2021 to investigate the possibilities of small modular nuclear reactors. Yet, as things stand, energy transitions in the four countries addressed in this book are more likely to follow the renewable energy path.

This book explores energy transitions from the standpoint of sustainability transitions – a research field with particular conceptual perspectives and approaches. It also draws from concepts in security studies to expand the outlook of sustainability transition studies. Energy transitions, together with energy and

---

[2] For example, the connections of nuclear power to security have been made via human and environmental security (Szulecki and Kusznir, 2018), terrorist attacks (Li et al., 2012), and military use of nuclear power (Johnstone and Stirling, 2020).

security policies in the case countries, are the empirical contexts in which these conceptual perspectives are applied.

Talking about security in the context of sustainability transitions may be tricky. I acknowledge that, with this book and the research it builds on, I contribute to "security talk" in energy transitions. This means that I am directing specific attention to security in connection with zero-carbon energy transitions and opening up transitions as a security question. However, initiating such security talk for sustainability transitions research does not amount to "securitization." Moreover, I argue that the issue is so important that it needs addressing. Further, I am not claiming that security trumps decarbonization. In fact, I take the opposite view – that decarbonization is vital and should be accelerated urgently but that the processes of accelerating transitions need to openly and critically consider how they influence different dimensions of security, so that societies are more prepared for these implications. There is also a positive dimension to security. Positive security can be thought of as the presence of conditions that further human wellbeing and promote peace (Hoogensen Gjørv, 2012; Roe, 2008). Such examples exist where renewable energy has been used to support peacebuilding efforts in different parts of the world. In turn, peace and prosperity, including the absence of armed conflict and promotion of human rights and social justice, are argued to be advanced "through accountable systems of governance and effective institutions of mature democracy" (Cortright et al., 2017, p. vii). This also supports the focus of this book on established and mature democracies.

So why does this book matter? First, when carrying out the early stages of this research, I encountered only a handful of people in each country who had expertise in both zero-carbon energy transitions and security. I hope that this expertise is continuously growing as transitions progress and, following the events of 2022, security has become a more pressing concern. The fifteen interviews conducted in each of the four countries in 2020–2021 probably covered almost all the expertise on this topic in those countries at the time of the interviews. Even many of those interviewed did not consider themselves experts but were all contributing some pieces to the puzzle. About a half of these experts were reinterviewed in 2022–2023, with some new experts added to the interview pool. Therefore, this book makes some of this rather limited knowledge accessible to more people and provides a holistic overview of security in connection to energy transitions.

Second, the scholarship on sustainability transitions has focused on innovations as a route to transitions and on the obstacles transitions may face. At the same time, the transitions themselves have mostly been seen as positive developments with little attention paid to their flipsides, the potentially negative developments arising from them. Some openings have been made, for example, in relation to mining (Marín and Goya, 2021). Yet, when I began writing this book, security was hardly addressed in this field of study. The exception was Phil Johnstone and colleagues, who were working on the military and the world wars in connection with sustainability transitions (Johnstone and McLeish, 2022; Johnstone and Stirling, 2020). Therefore, this book reveals new insights into sustainability transitions by focusing on security.

Third, research on the geopolitics of energy is often focused on mega states, such as the US, China, or Russia, or relatively large European countries, such as Germany or Poland, which can be seen as major geopolitical players. This book approaches the phenomenon of security in sustainable energy transitions from a small country perspective, of nations with circa five million inhabitants or less, where militaries are often focused on defence only. By doing so, the book may be of interest also to larger audiences regarding the different ways in which security plays a role in energy transitions. The small countries selected for the study are by no means similar or homogenous and provide different insights into the topic.

Next, I briefly introduce the conceptual background for this book, drawing from sustainability transition studies, security studies, geopolitics of energy, and policy studies, which are further elaborated in the following chapters.

## 1.1 Conceptual Background

This book is positioned in the sustainability transitions literature, which examines how sociotechnical change for societal service provision, such as energy, mobility, food, or water, is proceeding toward environmental sustainability and how such change can be better promoted. Integrating the security perspective into the transitions literature, the book addresses two points: first, how issues of security influence sustainability transitions and, second, what implications the transitions have for security in the energy sector context. The book also draws on policy studies, in particular the concepts of policy integration and policy coherence, to examine how energy, security, and defence policies are connected to each other, and together influence sustainability transitions.

The first domain of interest here, sustainability transition studies, has evolved since the late 1990s (Kemp et al., 1998; Rip and Kemp, 1998) as an interdisciplinary field of social science that looks at how large-scale transformations in systems for societal service provision occur and how such transitions can be promoted. It draws on the idea of *sociotechnical systems* that can be shaped by a dynamic interaction with disruptive niche innovations or broader landscape changes (Geels, 2004) and looks at how new sociotechnical systems emerge via technological innovation systems (Markard, 2020). In effect, the research field combines different types of frameworks – with historical, present, and future orientations – to examine sociotechnical transitions. The origins of this field drew substantially from science and technology studies, especially innovation studies, as well as evolutionary economics and historical studies. Over time, however, new perspectives have been added, for example, from geography, policy studies, and sociology. The field has developed into a substantial contribution to academic literature, with transitions scholars as authors in leading journals, such as *Research Policy* and *Global Environmental Change*, and on lists by Clarivate and Elsevier of highly cited academics. The number of publications on sustainability transitions has rapidly increased and accumulated, and the field's policy impact is visible in the European Commission and the Organization for Economic Co-operation and Development (OECD). Given the field's interdisciplinary nature, I argue that interconnections to security studies are also relevant when transitions accelerate in the real world.

Studying security is no longer just the prerogative of international relations scholars but of rising interest to scholars from other disciplines, such as law, criminology, anthropology, geography, and philosophy (Floyd, 2019). Thus, an application of security is also fitting for sustainability transitions research and is the second domain of interest in this book. The term security is used to refer to the absence of threats to or sufficient protection for acquired values (Booth, 1991; Wolfers, 1952), such as territorial integrity or political autonomy. Security is often also used in reference to peace, that is, the absence of armed conflicts or, more broadly, the presence of human rights and social justice (Cortright et al., 2017). Whereas, in the past, security may have been associated mainly with military threats and the protection of states, critical security studies challenged this conceptualization and opened up other security questions. These have concerned, for example, the environment, economy, or politics, and different reference objects (i.e., the objects to be secured); the field has also focused on the consequences of securitizing non-military issues (Peoples and Vaughan-Williams, 2015). The contemporary focus, further, examines security in relation to, for example, natural catastrophes, economic distress, shortages of essential supplies (e.g., food, water, energy), and people's everyday safety.

In this book, I make a distinction between negative and positive security following Roe (2008, 2012) and Hoogensen Gjørv (2012). Hoogensen Gjørv (2012, p. 836) has argued that "negative security can be understood as 'security from' (a threat) and positive security as 'security to' or enabling." Thus, for example, energy transitions can reduce or increase threats to the energy system, or the society at large, or they can improve wellbeing and, in this way, add to positive security. Indeed, positive security has also been connected to enabling individuals and communities (Booth, 1991, 2007). Positive security offers a different way to look at security (Hoogensen Gjørv and Bilgic, 2022). It can be increased, for example by providing social goods, such as education, healthcare, and public infrastructure, and, more generally, by means of good governance (Cortright et al., 2017).

Energy questions have been one aspect in security studies (Natorski and Herranz-Surralles, 2008). Moreover, energy security studies has become a rather extensive field in itself (Cherp and Jewell, 2011; Szulecki, 2018a). In addition, research on the geopolitics of energy is a particular field of international relations. This field has addressed the geopolitics of renewable energy (Scholten, 2018) and the geopolitics of hydrogen (Van de Graaf et al., 2020), which have become more important as energy transitions have progressed. Classical realist geopolitics is concerned about the ways in which *geographical factors* influence international relations. In contrast, critical geopolitics questions such straightforward investigation and is more interested in how *geographical assumptions* play a role in global politics. Literature on the geopolitics of renewables has expanded rapidly, focusing especially on questions such as the peace and conflict potential of renewable energy, potential winner and loser countries in the energy transition, and the consequences of renewable energy for international relations (Vakulchuk et al., 2020). Blondeel et al. (2021) highlighted the need to address the consequences of not only increased use of renewable energy but also hydrocarbon decline on geopolitics and international relations.

A third stream of research, which I draw from conceptually, are studies on policy coherence and integration, which have theorized different mechanisms of policy interplay. Since the late 1990s and early 2000s, there has been specific interest in these concepts in both environmental and development policy communities. Multiple empirical studies in these contexts have resulted in rather widely adopted conceptualizations of policy coherence and integration. In this book, I refer to policy integration as "the integration of a specific policy objective into another policy sub-system [or policy domain], for instance the integration of national security objectives into energy policy" (Kivimaa and Sivonen, 2021, p. 3). Policy coherence, in turn, is defined "as an attribute of policy that systematically reduces conflicts and promotes synergies between and within different policy areas to achieve the outcomes associated with jointly agreed policy objectives" (Nilsson et al., 2012, p. 396). The idea for both concepts is that solving complex policy dilemmas, such as the decarbonization of societies, requires connecting different policy domains and administrative sectors to reduce the number of conflicting incentives and rules given to stakeholders and to improve synergies. Specific processes in society are influenced by multiple administrative sectors. Different policy domains may inadvertently give conflicting messages to different actors who are meant to change their actions based on public policies, thereby reducing the effectiveness of these policies. Such policy incoherence may also be more costly to public administrations and transition pursuits.

Studies on policy coherence and integration offer specific frameworks to analyze the connections between and within different policy domains via different processes and levels of governance. Common analytical dimensions include, for example, horizontal, vertical, internal, and multilateral coherence (Carbone, 2008). Integration, by contrast, has been divided into four types of approaches: normative approaches that emphasize the principled priority of environmental or climate change issues; organizational and procedural approaches that propose mechanisms to deliver policy integration; output-based assessments of the achievement of policy integration; and approaches focused on learning and reframing (Russel et al., 2018). This book is interested in how the dynamics of policy coherence and integration play out in the context of energy transition policy and national security and defence policy.

## 1.2 A Small Country Perspective: Estonia, Finland, Norway, and Scotland as the Research Foci

As mentioned, in contrast to previous studies addressing energy security from the perspective of major states or larger countries, this book is interested in how smaller countries approach the interface of zero-carbon energy transitions and security. The focus is on the nation-state level because states have historically been, and still are, the entities responsible for energy infrastructure, security, and public policy.

The countries addressed in this book are geographically located in Northern Europe and have populations of around five million people, or less. They are also all part of the same interconnected electricity network that spans the Nordic and Baltic countries, with the connection between Norway and the UK opened most recently. The North Sea link – over 700 kilometers (450 miles) long – became operational in October 2021.

The countries also show a lot of variation, for example, in their energy profiles and their stance to security and policymaking. Estonia, Norway, and Scotland (as part of the UK) are longtime members of the North Atlantic Treaty Organization (NATO), while Finland became a member in 2023. Finland and Estonia are member states of the EU. Estonia, Finland, and Norway share a border with the energy and military superpower Russia, but, as the analysis in this book shows later, this has had very different effects on energy policymaking in each country. This is because the countries' stances on hydrocarbon phaseout, their import dependencies, and geopolitical positionings have differed as has their history with Russia.

In Estonia, oil shale has been an important domestic energy source, providing energy independence and employment. The climate change mitigation policies of the EU have challenged the Estonian energy system and thus Estonia has faced the need to phase out oil shale and expand its wind power sector. The latter has created a complicated situation regarding the defence of the country, because high wind turbines near the Russian border interfere with the operation of air surveillance radars and signal intelligence. Further, Estonia, alongside the other Baltic States, is seeking to desynchronize its electricity network from Russia by 2025, if not sooner. Estonia, alongside other post-Soviet countries, has considered security to be part of energy policy for a long time. Therefore, the analysis of this country helps in understanding differences in national energy transitions across Europe, given the strong contrast between Estonia's situation and those of my other case countries. Chapter 5 goes into these topics in more detail.

Finland has a very different energy profile to Estonia because it has no domestic oil, coal, or gas reserves, and its share of peat (a carbon-intensive fuel) is small. It has a high share of renewable energy in its mix but faces still a challenging task in moving its heating and transport sectors away from imported fossil fuels and increasing the electrification of society with the help of wind power and the new nuclear reactor that began operating in 2023. Finland has experienced a similar debate as Estonia regarding the interference of wind power with the operation of their national air surveillance systems. Yet, prior to 2022, it much less explicitly connected energy to national security than Estonia, and energy was intentionally desecuritized (see details in Chapter 6).

Norway differs from these two countries by being completely self-sufficient in energy thanks to its large hydropower reserves, as well as extensive hydrocarbon production, which is mostly exported. In addition, Norway has been the leading country worldwide in electrifying transport. Wind power has been under much debate, because the intention has been to provide electricity for the European market, a proposal not liked by all Norwegians. Security was not a big question prior to 2022 and was related mostly to: (1) economic security following calls for hydrocarbon phaseout, because oil has been such a large source of income for Norway; and (2) the Norway–Russia dialogue pertaining to oil exploration in the High North. Chapter 7 describes the Norwegian case.

Scotland is a special case in this study as it is not an independent nation but a constituent nation of the UK. While Scotland has its own policies, for example regarding energy efficiency and spatial planning, it is dependent on decision-making concerning security and energy policy by the UK parliament. The independence debate in 2016 and related documentation revealed Scotland's own ambitions for energy in terms of increasing renewable energy and avoiding nuclear power, but also the challenges of phasing out hydrocarbons. The security questions around the energy transition pertaining to Scotland are multifaceted due to the country being an integral part of UK electricity and gas networks as well as the fact that UK nuclear submarines have their base on the Scottish coast (see details in Chapter 8).

## 1.3 Research Method and Materials

When writing this book, I sourced different materials, including eighty-eight in-depth interviews in Estonia, Finland, Norway, and Scotland/the UK; over seventy policy strategy documents published during 2006–2023, the period of focus in this study; and a range of secondary materials such as media articles and reports.[3] The interviews were conducted in two separate rounds and in a semi-structured manner, where the same set of main questions was followed by supplementary questions added based on the interviewees' backgrounds. The interviewees were identified as having expertise at the cross-section of energy, security, defence, or international affairs. Therefore, they had complementary skills to each other even if their position leant toward a particular direction. The interviews, conducted in English or Finnish, were recorded and transcribed for analytical purposes. The first round of sixty-one interviews, with sixty-six interviewees, took place between September 2020 and May 2021 in the form of online video calls. Five interviews comprised more than one person placed in the same organization. The interview durations ranged from 28 to 107 minutes, on average lasting 69 minutes. The second round of interviews was smaller, with an aim to reinterview around half of those who had been involved in the first round, to get a sense of changes in thinking and policy development since the 2022 events. The second round also included six completely new interviewees due to the unavailability of some of the previous experts. The structure of the second-round interviews was updated based on contemporary developments. The second round comprised twenty-seven interviews, with thirty-two interviewees, conducted between November 24, 2022 and March 6, 2023. The interviews lasted from 26 to 78 minutes, on average 56 minutes. Three interviewees included more than one person from the same organization.

Table 1.1 shows the division of interviewees based on their country and primary affiliation, noting that some interviewees had worked in both the private and public sectors or academia and the public sector. Around two-thirds of the interviewees had energy system or energy policy-related expertise. A third were experts in security and defence. A sixth had expertise in international relations, and a similar number were involved in party politics. The interviewees worked in ministries for economic affairs (in charge of energy policy), foreign affairs, and defence. In addition, the interviews covered people working for research institutes (especially on international affairs) and universities, government agencies, transmission network operators, and energy companies, or those who were members of national or EU parliaments.

The interview material is rich and brings forth diverse issues. These include, for example, the influence of energy transitions on security questions that range

---

[3] The interview and document material were collected jointly with Marja H. Sivonen.

## 1.3 Research Method and Materials

Table 1.1   *Principal affiliations of the interviewees*

|  | Public sector (ministries, agencies) | Private sector (e.g., energy companies) | Research (universities, research institutes) | Politics (members of national/EU parliaments) |
|---|---|---|---|---|
| Estonia (*n* = 19) | 9 | 2 | 6 | 2 |
| Finland (*n* = 19) | 8 | 4 | 4 | 3 |
| Norway (*n* = 19) | 7 | 5 | 5 | 2 |
| Scotland (*n* = 15) | 5 | 1 | 6 | 3 |
| **Total (*n* = 72)** | **29** | **12** | **21** | **10** |

from energy security via geopolitics to cybersecurity; the interplay between different ministries involved in energy and security questions; the worldviews guiding policymaking and public administration; the policy developments taking place; and the influence of Russia on the countries' energy sectors and policymaking.

Policy strategy documents from the period 2006–2023 were, furthermore, extensive, at around 8,000 pages (Table 1.2). Examples include the "National Development Plans for the Energy Sector" and "National Security Concept" publications in Estonia, the "National Energy and Climate Strategies" and "Government Reports on Security and Defence Policy" in Finland, "National Climate Policy" and "Cybersecurity Strategies" in Norway, and the "Scottish Energy Strategy" and the "National Security Reviews" in Scotland/the UK. The paragraphs in the documents that connected energy policy to security questions or security policy to energy questions were identified and coded in the NVivo software program. While some policy documents made rather broad connections between energy and security, others hardly addressed it. More detailed document analyses conducted in the project that this book draws from are reported elsewhere (Kivimaa and Sivonen, 2021; Sivonen and Kivimaa, 2023), but they are also referred to in this book's country analyses (Chapters 5–8).

During the study, the material was processed and analyzed in different ways. For example, the materials were systematically analyzed using the software tool NVivo and the spreadsheet program Excel for qualitative analysis. However, in this book, a broader approach was also taken by putting this material together in writing the country chapters and telling the story of how zero-carbon energy transitions connect with security questions in each country. Thus, the resulting book is a mix of inductive and deductive analyses of the materials collected for the research project, generating a summary of the topic and insights from the past and for the future.

Table 1.2  *The number of policy documents studied*

|  | 2006–2010 | 2011–2015 | 2016–2020 | 2021–2023 |
|---|---|---|---|---|
| Estonia | 4 energy/climate, 3 security/defence documents | 1 energy/climate, 3 security/defence documents | 3 energy/climate, 4 security/defence documents | 0 energy/climate, 1 security/defence documents |
| Finland | 2 energy/climate, 3 security/defence documents | 3 energy/climate, 4 security/defence documents | 3 energy/climate, 6 security/defence documents | 1 energy/climate, 1 security/defence documents |
| Norway | 2 energy/climate, 4 security/defence documents | 4 energy/climate, 6 security/defence documents | 6 energy/climate, 9 security/defence documents | 1 energy/climate, 0 security/defence documents |
| Scotland | 5 energy/climate, 2 security/defence documents | 5 energy/climate, 7 security/defence documents | 6 energy/climate, 6 security/defence documents | 2 energy/climate, 3 security/defence documents |
| Total | **13 energy/climate, 12 security/defence documents** | **13 energy/climate, 20 security/defence documents** | **18 energy/climate, 25 security/defence documents** | **4 energy/climate, 5 security/defence documents** |

## 1.4  Contents of the Book

This book is divided into three parts. It starts, in Part I, with the theoretical and conceptual foundations of connecting sustainability transitions and security with each other. This part contains Chapter 2, which introduces research and key concepts in the field of sustainability transitions and moves into introducing security studies and its central concepts. It also discusses how security has thus far been addressed in the field of sustainability transitions and explores what conceptual developments are needed to address security more thoroughly among other side effects of sustainability transitions. Chapter 3 reviews previous research on energy security and the geopolitics of renewable energy as an important context in which research on the interconnections of sustainability transitions and security is placed. Chapter 4 presents the conceptual–analytical framework to examine the country cases, drawing from sustainability transition studies, security studies, and the literature on policy coherence and integration. It also explains the concepts of policy coherence and integration and reviews relevant literature, some of which ties policy interplay into the literature of sustainability transitions.

Part II contains four country-specific analyses of the connections between zero-carbon energy transitions and security, and the domains of energy policy and security and defence policies. Chapter 5 focuses on Estonia (energy independence and oil shale), Chapter 6 is about Finland (ambivalent links between energy and security policy), Chapter 7 looks at Norway (contradiction of oil and renewables for economic security), and Chapter 8 is about Scotland (from oil to wind under devolved government). My research team members have provided contextual material for some of these chapters. Chapter 5 draws partly from a background document produced by Claire Mosoni on Estonia, and Chapter 7 on Norway benefits from background work and insights provided by Marja H. Sivonen.

Part III consolidates the theoretical foundations and the country cases to generate new insights for the academic study of sustainability and energy transitions, as well as for policymakers and other experts interested in the topic. Chapter 9 compares the country findings and brings together the conceptual and empirical insights presented. It first discusses the interplay between energy, security, and defence policies, followed by securitization and politicization. Subsequently, focus is placed on the security implications of energy transitions, and negative and positive security. The chapter ends by summarizing the key technological, actor-based, and institutional aspects of the country cases, Russia as a landscape pressure, and final conclusions.

One of the things that this book aims to uncover is whether there is or has been a hidden side to policymaking that considers energy questions from the perspective of national security, and how this hidden side could have influenced (or may in the future influence) the achievement of sustainability transitions. It also discusses whether the security implications of energy transitions are similar or different under centralized versus decentralized renewable energy transition pathways, paying attention to both negative and positive security. It is, nevertheless, clear that, with or without the energy transition, the security issues connected to the energy system will change. The world is facing an increasing demand for energy, a further electrification of energy systems, and new types of cyber and hybrid threats, which will shape how energy and security policies evolve and need to interconnect. Further, the rise of the far right and the rising contradictions between the thinking of the far right and liberal environmentalists pose a different kind of risk for societal stability. The global energy transition (or the lack of it) is connected to domestic stability within countries as well as to international relations and global stability. This means that new kinds of interactions and coordination efforts are needed among policy domains responsible for climate change, energy, international relations, security, and defence. In addition, a redefinition of what energy security and security of supply mean in this new context may be necessary.

# Part I

Theoretical and Literature-Based Foundations

# 2
# Understanding Security in the Context of Sustainability Transitions

The field of sustainability transitions is relatively new in academic research, with the first publications on the subject appearing in the late 1990s. It has witnessed a rapid expansion since that time, both in terms of number of publications and empirical and conceptual reach. This field can be characterized as having a multi-disciplinary social science orientation; extensive conceptual development; a normative focus on real-world environmental (and social) problems; and a transversal and multiscalar approach. Summaries of the field thus far provide a more detailed account of the key features and developments (Köhler et al., 2019; Markard et al., 2012; Truffer et al., 2022).

The term "sustainability transition" refers to whole-system changes and complete reconfigurations of the combinations of technologies, infrastructures, practices, and institutional structures that have formed around the provision of societal services – such as energy, transport, food and agriculture, and water – and industrial production. The energy transition, for example, implies not only substantial change in the technologies used for energy production and consumption but also institutional shifts in energy markets and regulatory structures as well as the practices of producing and consuming energy (Johnstone et al., 2020). Therefore, energy saving has an important role to play in achieving whole-system energy transition, although this aspect is often ignored, perhaps due to its low-tech nature and the fact it inspires little interest within political decision-making.

Sustainability transitions research began with an orientation into how (socio) technical innovations can improve environmental conditions and reduce environmental pollution on the planet by infiltrating dominant sociotechnical systems and substantially changing them. This orientation was complemented by the development of conceptual frameworks to study these processes. After over two decades of development of this field, and as sustainability transitions have begun to advance more concretely in real life, new advances and concerns have become part of transitions research. These include, for example, the broader repercussions of sustainability transitions (Kanger et al., 2020), considering and alleviating the impacts of transitions on social justice (Kaljonen et al., 2021; Sovacool et al., 2019), and the dynamics at play when transitions cross sectoral boundaries (Geels, 2007; Schot and Kanger, 2018).

This chapter introduces key conceptualizations of sustainability transitions research, and then takes a particular focus on security. The chapter highlights some key transitions terminology and processes that are relevant for discussing the empirical context of energy transitions and security and unpacks the basics of security. It also reviews the limited transitions literature that addresses aspects of security, excluding my work on this topic, which is covered later in the book.

## 2.1 Sustainability Transitions Research: Key Conceptualizations

Sustainability transitions research began with a focus on a set of processes that could result in fundamental shifts in sociotechnical systems. These processes are associated with extensive adjustments to technological, material, organizational, institutional, political, economic, and sociocultural elements (Markard et al., 2012). *Sociotechnical* is the guiding perspective in sustainability transitions research. The key factors of sociotechnical systems are regarded as technologies, actors, and institutions.

*Technology* refers to material or virtual artifacts and knowledge, ranging from minor technical components to entire economic sectors (Kivimaa et al., 2022). In transitions studies, it is understood "with respect to a function embedded in a reasonably complex focal product," such as "a wind turbine that converts wind to electricity" (Andersson et al., 2021, p. 113). Technology has been and is a key focus of transition studies, especially during the first decade of the field's development.

*Actors* comprise those who have specific roles in the established sociotechnical system and in advocating new niche technologies or services. They can be individuals, organizations, networks of individuals and organizations, or even state governments. Some actors advance transitions, while others may actively oppose them. The actor dimension is connected to the power to advance or hinder things (i.e., "power to"), dependencies between actors (i.e., "power over"), and the power of coalitions of actors ("power with") (Avelino, 2021). Thus, transitions are also about shifting power relations between actors (Avelino and Wittmayer, 2015). Actors affecting energy transitions comprise, for example, energy producers and consumers; transmission and distribution network operators; public agencies and officials regulating energy production and supply; scientists and innovators developing new technologies and services, and others who advance them; as well as nongovernmental organizations (NGOs), and individual citizens, residents, and Indigenous communities, who influence and are impacted by energy transitions.

*Institutions* regulate and guide actors' actions and relations, for example, by fear of sanctions or by shaping beliefs or values (Geels et al., 2016; Ghosh and Schot, 2019). Institutions include (semi)permanent formal and informal rules, regulations, standards, and social norms. Regulatory, normative, and cognitive institutions (cf. Scott, 2001) are embedded in sociotechnical systems as the rules that form their deep structure (i.e., regimes) and guide actor perceptions, behavior, and activities (Geels, 2004, 2006; Geels and Schot, 2007). Shared and stabilized rules between regime actors constitute a *sociotechnical regime* manifested in different dimensions, such as market and industry structure, public policy and politics, and symbolic meanings associated with culture. Therefore, sustainability transitions are about changing the underlying rules of the system, not simply the system's technical configurations (Ghosh et al., 2021). Actors can join together to construct supportive institutional structures around new technologies and services (Musiolik et al., 2012).

The early literature on sustainability transitions built four conceptual approaches that all share an orientation in terms of the "sociotechnical" and a normative purpose to advance environmentally sounder transitions. Otherwise, the approaches have somewhat differing starting points and theoretical or empirical influences. The approaches include the multilevel perspective (MLP), strategic niche management (SNM), transition management (TM), and technological innovation systems (TIS) (see, e.g., Markard et al., 2012, for details). This book mainly draws on ideas from the MLP and SNM, but also more generally from broadening transitions research. The research field has fluid boundaries, informed by the shared normative orientation of these four conceptualizations (Kivimaa et al., 2019). It is, moreover, increasingly widening and becoming more open to new conceptualizations from an increasing number of social science fields.

Influenced by the MLP and SNM, "niches" and "regimes" are widely used central concepts in transition studies, albeit not applied in all approaches (e.g., TIS). *Niches* are described as spaces for experimentation and radical innovations, while *sociotechnical regimes* are fairly stable, shared, and dominant configurations of technologies, institutions (i.e., rules), practices, and actor networks (Geels, 2002; Rip and Kemp, 1998).

*Landscape* is a somewhat less used concept, associated only with the MLP. It is seen as the selection environment for niches and regimes, determining the conditions for their operation and exposing them to new pressures once these conditions change (Berkhout et al., 2009). It is a slow-moving and relatively stable heterogeneous collection of issues, such as environmental problems and globalization (Grin et al., 2010), or, more broadly, political and sociocultural contexts (Berkhout et al., 2009). It was first introduced by Arie Rip and René Kemp, in association with technological change, as the social context into which new technologies are presented; they also suggested that technologies can contribute to the sociotechnical landscape, and provide an example of motorcars influencing broader ideas of freedom and democracy (Rip and Kemp 1998). Geels (2011, p. 27) posits that landscape is a derived concept because it is always defined in relation to the sociotechnical regime "as [an] external environment that influences interaction between niche(s) and regime." Sometimes landscape has been criticized as being a "garbage can" of contextual influences (Geels and Schot, 2007) and difficult to operationalize in practice (Rock et al., 2009). Yet, more recent research has created openings, with a more structured understanding of landscapes (e.g., Antadze and McGowan, 2017; Morone et al., 2016).

A process orientation is central in transitions research. One of the core notions in the unfolding of sustainability transitions is *coevolution*. Analysis of coevolution aims to detect causal interactions between evolving systems or subsystems (Foxon, 2011). Coevolution implies a situation in which two or more (sub)systems are connected so that each affects the development of another (Safarzyńska et al., 2012). In the context of sociotechnical systems change, this refers to the dynamics of change between economic, cultural, technological, ecological, and institutional subsystems that influence the speed and direction of transitions (Grin et al., 2010). The transitions literature also uses process orientation in the more specific context of a coevolving mix of policies with the sociotechnical regime. In this specific context, the sociotechnical regime creates political, administrative, and fiscal feedback to the development of policy mixes, and the changing policy mix affects the resources, interpretations, and institutions of the sociotechnical regime (Edmondson et al., 2019).

Coevolution is also behind the MLP, which describes transitions as resulting from interplay between the three levels – niches, regimes, and landscape. Initially, innovations deviating from the regime are developed as small initiatives in the niches, which can grow larger and break through to the regime level; the success of the breakthrough being dependent on the pressures that the landscape level puts on the regime (Geels, 2005a). A typology of transition pathways identifies differing dynamics between niches, regimes, and the landscape, depending on the timing of interactions by which this coevolution occurs. Geels and Schot (2007) describe *transformation* as a process, where moderate disruptive landscape pressure occurs at a time when niche innovations are not sufficiently developed, and where regime actors respond by guiding the direction of innovation activities. If landscape change is sudden, large, and creates problems for the regime, a *dealignment* of the regime creates space for several niche innovations, one of which eventually gains momentum and becomes *realigned* to a new regime. A third dynamics is *technological substitution* where substantial landscape pressure (a long-term disruption or a sudden shock) happens when niche innovations are well developed but have been unable to break the regime in the past. We can see elements of dealignment/ realignment and technological substitution in the European energy sector's reactions to Russia's invasion of Ukraine in 2022. Finally, *reconfiguration* is described by Geels and Schot as symbiotic niche innovations initially adopted at the regime level to solve local problems, but as triggering further adjustments in the basic regime configuration, resulting in more substantial changes than envisaged.

Subsequently, both transformation and reconfiguration have been defined in multiple ways. Reconfiguration has, in contrast to the Geels and Schot definition, been viewed via the lens of a whole systems approach. It is "used to illustrate how the hierarchies between the niche-regime-landscape relations are becoming blurred and questioned" (Laakso et al., 2020, p. 16). Transformation toward sustainability is used as a concept beyond sustainability transition studies. It is generally understood somewhat similarly to sustainability transitions: as a significant reordering that challenges existing structures to produce fundamental novelty (Blythe et al., 2018).

SNM, which is connected to the MLP, has paid much attention to processes by which niches are created and how they may accelerate and become institutionalized as part of new sociotechnical regimes (e.g., Raven et al., 2010; Schot and Geels, 2008; Van der Laak et al., 2007). In a seminal article, Smith and Raven (2012) described shielding, nurturing, and empowering as key contributions to wider transition processes. Shielding refers to processes that create conditions for niche innovations to develop by protecting them from incumbent interests (Ghosh et al., 2021) and the mainstream selection environment of the sociotechnical regime (Smith and Raven, 2012). Nurturing of niche innovations is articulated as three intertwined processes that take place within a protective space (i.e., the niche) (Hoogma et al., 2002; Schot and Geels, 2008): (1) articulating expectations and visions shown via multiple experimental projects and shared by actors; (2) creating and managing networks where niche actors cooperate and combine resources; and (3) learning in multiple dimensions, aggregating knowledge from experiments, and sharing it forward. Table 2.1 describes these processes in more detail.

Recent research has begun to devote more attention to the processes of accelerating, embedding, and institutionalizing niche innovations. This is a natural follow-up to some advancing real-world processes, and it shows both the need for and progress in accelerating promising niche innovations. The literature posits, for example, that, for accelerating transitions, public policies need to shift from supporting individual innovations to a wider system-change approach and to better acknowledging multisystem interactions (Markard et al., 2020). A system-oriented approach and multisystem interactions connect to the horizontal policy-coherence perspective taken in this book (see Chapter 4).

Table 2.1   Processes of SNM

| Niche development process | Grounding in literature |
| --- | --- |
| Articulating expectations and visions | Various actors engage in niche-building processes, and separate expectations shape into niche actors' shared expectations about future developments and shocks at the landscape level, how sociotechnical regimes will respond to these, and what kind of potential niche innovations offer. These expectations can be unpredictable. Expectations guide learning processes and gain attention from more actors and resources. This process is productive if actors start having similar expectations, and if expectations become more specific (Ghosh et al., 2021; Schot and Geels, 2008; Van der Laak et al., 2007). |
| Building social networks | In the preliminary stages of niche development, social networks are feeble and transitions depend on the collaboration of numerous actors. Networks are formed to create a community behind the niche by enabling interactions and allocating resources. The process is successful if networks are broad and oriented toward deep learning, and if regular interaction is supported (Ghosh et al., 2021; Schot and Geels, 2008; Van der Laak et al., 2007). |
| Learning | Niche development occurs via various forms of learning, for example, technical, market, cultural, and policy learning supported by several experiments. Learning can be described as a perceptive process of knowing, understanding, and reflecting. Deeper learning, which moves from gathering facts and data to changing cognitive frames and assumptions, is important. The process is successful if it combines technological change with societal embedding in local contexts and addresses multiple dimensions (Ghosh et al., 2021; Schot and Geels, 2008; Van der Laak et al., 2007). |

*Source:* Adapted from Kivimaa and Sivonen (2023).

Geels and Schot (2007) have suggested indicators to recognize when niche innovations may be ready to be diffused more widely: first, learning processes in the niche have stabilized into a reasonably dominant design for the innovation; second, powerful actors have joined the support network; third, the price–performance ratio has improved, with strong expectations for future advancement; and, fourth, market niches for the innovation amount to more than 5 percent of market share. A good example of an innovation meeting these indicators is wind power technology.

The acceleration of niche innovations connects to a process of empowerment. Smith and Raven (2012) argue that once niche innovations have become competitive within a conventional regime context, protective shielding becomes redundant and the innovation is empowered and able to be diffused more widely. However, this kind of empowerment does not necessarily mean that niche innovations will accelerate in a way that alters incumbent sociotechnical regimes substantially. Smith and Raven (2012, p. 1030) distinguish fit-and-conform empowerment, where niche innovations become competitive within unchanged selection environments, from stretch-and-transform empowerment, where "some features of the niche space are institutionalized as new norms and routines in a transformed regime." An example of fit-and-conform empowerment is when biofuels were added to transport fuels without substantially changing the transport regime. Contrarily, solar photovoltaics have stretched and transformed the energy regime in many localities, allowing a more distributed production of electricity, as well as enabling consumers to act as both producers and consumers of electricity.

The literature on acceleration is focused on processes to characterize different forms of niche expansion and embedding, with the aim of depicting how niche experiments or completed niche innovations diffuse and their broader transformative impacts. Naber et al. (2017) proposed a typology of patterns for expanding transition experiments. First, they identified *growing* as either an increase in the number of participants in the experimentation or an increase in the degree to which a new technology is used. Others use the term *upscaling* to describe a similar process (Ghosh et al., 2021; Gorissen et al., 2018). However, Turnheim et al. (2018) proposed that besides expanding the scope and length of an experiment, upscaling can also, for example, be about mainstreaming knowledge and learning or about new practices generated during an experiment. *Replication* was proposed by Naber et al. as an application of the main concept of the experiment in other contexts. Again, a more nuanced interpretation was proposed by Turnheim et al., where replication may also involve the experiment's actor configuration, the technology or service provided, or the diffusion and recontextualization of knowledge. Accumulation is seen as a process where experiments are linked to other initiatives (Naber et al., 2017). This can be connected with *circulation*, which is about the flow of ideas, people, or technologies between experiments or niches (Ghosh et al., 2021; Turnheim et al., 2018). Finally, *institutionalization* is a process where the experiment or niche shapes the regime selection environment (Naber et al., 2017). Knowledge and learning generated in experiments become new rules, practices, and scripts; policy outputs or practices become embedded in formal and informal governance structures; and technologies and services become widely adopted (Ghosh et al., 2021; Turnheim et al., 2018).

The ways in which dominant and established sociotechnical regimes decline to make space for transitioned regimes was hardly discussed in early transitions research. Indeed, many have argued that transitions studies suffer from an "innovation bias" that exaggerates novelty at the cost of undertheorizing decline (Feola et al., 2021). Recently, more attention has been devoted to decline via various conceptualizations of destabilization (Koretsky et al., 2022; Turnheim and Geels, 2012) and phaseout (Isoaho and Markard, 2020; Rogge and Johnstone, 2017). The research on phaseout has typically oriented to technological decline and discourses, while the destabilization literature has adopted a whole systems perspective (Kivimaa and Sivonen, 2023).

Feola et al. (2021) argue that niche creation is always coupled with a disruptive side, where experimentation meets resistance and propositions are refused. This links to the argument that, essentially, destabilization or decline are necessary conditions for transitions (Kivimaa and Kern, 2016), while "transition" itself can range from disruption (Kivimaa et al., 2021) to more subtle reconfiguration (Laakso et al., 2020). The processes related to decline are likely to face resistance, opposition, and tensions, which may, in extreme cases, become adverse security consequences, such as physical conflicts or riots. Further, in connection to disruptions, the role of actors in transitions has received increasing attention and the old dichotomy between niche and regime actors is being replaced with more nuanced insights. Incumbency is no longer characterized merely as no or slow action; it has been recognized that there are variations in how incumbent actors react to transitions (Sovacool et al., 2020; Stirling, 2019).

The literature on niche development, combined with the idea of destabilizing or dealigning sociotechnical regimes, has led to the development of "transformative outcomes," which describe processes that transition actions should aim to promote. These transformative outcomes, with an aim to "lead to deeper changes in sets of rules that guide actors," are built around three macro-processes of transitions: building and nurturing niches; expanding and mainstreaming niches; and unlocking and opening up regimes (Ghosh et al., 2021, p. 741). These are partly sequential in that expanding and mainstreaming cannot occur before building and nurturing, while the unlocking and opening of regimes can happen in parallel. One of the key associations of this is that how security connects to sustainability transitions looks different in a relatively early phase when new niches are being developed, as opposed to when niches expand and become mainstream, or, especially, when established regimes destabilize. This is further addressed in Chapter 4, which outlines the analytical framework adopted in this book.

Finally, I want to remark that since 2020, transition scholars' interest in the broader repercussions of sustainability transitions has expanded. There is recognition that sustainability transitions are not all about positives but have potentially negative side-effects while transitions unfold. Some of these effects are limited to the duration of the major shift, while others may prevail in the new regimes. Kanger et al. (2020) mention broader repercussions and refer, for instance, to the need for policies to anticipate and alleviate transitions' unintended consequences. Some of these consequences relate to injustice and inequalities that transitions may create and, hence, many policy efforts are directed at just transitions. Ghosh et al. (2021, p. 741) argue that transitions "involve addressing systemic inequality, injustice, and marginalization of actor groups, including unequal distribution of benefits." Nonetheless, neither security nor reduced security have been mentioned among these broader repercussions and these require further attention.

## 2.2 Conceptualizing the Basics of Security for Sustainability Transitions

Security studies have been argued to be the most widely studied subfield of international relations (Floyd, 2019). Early on, in this context, security was particularly associated with military threats and the protection of states, so the traditional, realist definition of national security was adopted; since then, security has developed into a contested concept for which a variety of meanings exist (Peoples and Vaughan-Williams, 2015). Buzan et al. (1998) described military security as the ability of governments to maintain themselves against internal and external military threats and the use of military power against nonmilitary threats to existence. Huysmans (1998), however, claimed that "security" does not refer to any external objective reality, but, rather, the term establishes the situation. Since the end of the Cold War, the concept of security has been broadened to many other contexts, such as climate change and human security. Nowadays, many governments, when they talk about national security, refer not only to military security but to multiple other things. The Finnish "Government Report on Foreign and Security Policy," for example, mentions a "comprehensive security" approach that acknowledges threats against societal well-being from hybrid influencing, climate change, and pandemics (MoFA, 2020).

One of the key terms in security studies is the "referent object." It means "that which is to be secured" (Peoples and Vaughan-Williams, 2015, p. 4). Traditionally, a key referent object would have been the state. Over time, the conceptualization of security has broadened and deepened, particularly when critical security studies challenged the idea that security should be understood solely in terms of military threats to the state (Peoples and Vaughan-Williams, 2015). Broadening refers to adding new sectors under the analysis of security. Such sectors as energy (Cherp and Jewell, 2011, 2014), food (Prosekov and Ivanova, 2018), the environment (Allenby, 2016), and water (Cook and Bakker, 2012) have been covered under the name of security studies. While security is much more than the absence of military conflict, Floyd (2019) criticizes the broadening literature on security for its nonspecificity regarding why security is valuable as a unit of analysis to these sectors. Environmental security, in particular, links to sustainability transitions, but energy, water, and food security also become relevant, because sustainability transitions aim to change sociotechnical systems built around these areas – with these changes also impacting their security context.

Deepening means that new referent objects have been added to security studies besides the state. Ecosystems and the natural environment became referent objects for security in public policy, media, and academic settings in the late 1980s (Peoples and Vaughan-Williams, 2015). In 1987, the United Nations published the "Report of the World Commission on Environment and Development: Our Common Future," under the aegis of Gro Harlem Brundtland, which led to worldwide attention being paid to environmental problems. Aiming to securitize the environment, the report referred to security 122 times and stated, for instance, that the "deepening and widening environmental crisis presents a threat to national security – and even survival – that may be greater than well-armed, ill-disposed neighbors and unfriendly alliances" (UN, 1987, p. 23). The report also focused on food security. Since then, environmental security has become its own distinct scholarly subfield (e.g., Dalby, 2002; Trombetta, 2009).

Another deepening of referent objects is humans. The origin of human security goes back to the UN development report in 1994 that recognized the human as a referent object for security (Peoples and Vaughan-Williams, 2015). The report mentioned security over 300 times, referring to "safety from the constant threats of hunger, disease, crime and repression … and protection from sudden and hurtful disruptions in the pattern of our daily lives – whether in our homes, in our jobs, in our communities or in our environment" (UN, 1994, p. 3). Hoogensen Gjørv states that "[h]uman security focuses upon the individual instead of the state as the security referent, which makes the approach appealing for its recognition of individual, 'everyday' security concerns, making individuals relevant and visible, and listening to marginalised voices" (Hoogensen Gjørv, 2012, p. 838). Objective existential threats to human security not only refer to lethal things but also those that threaten basic human needs to live minimally decent lives, such as disabling infectious diseases (Floyd, 2019). Therefore, health pandemics are existential threats to human security alongside the implications of climate change, such as the flooding of human settlements or extreme temperatures. The human security dimension connects to the rapidly growing literature on just sustainability transitions (see Kaljonen et al., 2021; Sovacool et al., 2019) and the actor dimension of sociotechnical systems.

*Securitization* was initially suggested by the Copenhagen School of International Relations as a tool to analyze security and the consequences the use of the term security has for nonmilitary issues or sectors (Peoples and Vaughan-Williams, 2015). It has often been seen as a negative process or phenomenon. Buzan et al. (1998) expressed that, when states or other actors securitize an issue, this is a political act that has consequences. Securitization is emphasized

## 2.2 Conceptualizing the Basics of Security

as a speech act with political implications (Hansen, 2012). Floyd (2019, p. 71) defines securitization as a process whereby "an issue is moved from normal politics to the realm of security politics, where it is addressed by security measures." She further argues that security politics are very different from ordinary politics due to the nature of decision-making; almost always resulting in some negative consequences and, at a minimum, a reduction of democracy. However, this perspective of securitization ignores those countries where democratic decision-making is not the norm (Aradau, 2004). Further, drawing from critical security studies and the Welsh School of Security Studies (Emancipatory Realism), Hoogensen Gjørv (2012) claims that if securitization has a "good" result it can be an example of positive security, although determining "good" is not always easy. This book adopts the latter, more nuanced, Welsh School interpretation of securitization.

According to the Copenhagen School, securitization is connected to an *existential threat* being present or expected. Although some security threats are socially or politically constructed, both Wæver (2011) and Floyd (2019) recognize the presence of objectively pre-existing threats that exist even when they are not labelled as such. Thus, security is not purely a social construction. According to Buzan et al. (1998), an existential threat requires emergency measures and justifies actions outside normal political procedures. Some security threats are caused by human agency, while others can be defined, according to Floyd (2019), as agent-lacking threats. The latter threats might occur, for example, as a result of natural disasters – including those caused by climate change – while, indirectly, they may also be induced by human agency. The securitization theory proposes that three steps are required for securitization to occur. First, a *securitizing move* is a discourse or a speech act where something is presented by a securitizing actor as an existential threat. Second, an *audience* needs to accept this claim; and, third, legitimize (albeit not necessarily adopt) *extraordinary emergency measures* in response (Buzan et al., 1998). Floyd (2019, p. 40) argues that "securitization is morally permissible only in the presence of an objective existential threat." It can, however, be argued that the Copenhagen School is vague about the difference between normal politics and extraordinary measures and who the audience is in the latter case (Heinrich and Szulecki, 2018).

Securitization and desecuritization partly arose as a counterargument to the broadening and deepening of the security concept. In the early 1990s, Ole Wæver, a central figure in the Copenhagen School, argued that security needs to be thought in terms of national security (and not in a broader sense), and that the dynamics of securitization and desecuritization cannot be analyzed if security is assumed to have positive value (Wæver, 1995). Desecuritization has since been used to mean a process whereby issues are shifted out of the "emergency mode" and into the sphere of normal politics, although the term has also been heavily criticized by many security scholars for being underspecified (Aradau, 2004; Hansen, 2012). Floyd (2019) argues that desecuritization as a process can result in formerly securitized issues being either politicized or depoliticized. Depoliticization has been described as "placing the political character of decision making at one remove from the central state" and delegating "decisions that are usually the responsibility of ministers … to quasi-public bodies that either advise or implement those political decisions, or [where] rules are created constraining ministerial discretion" (Wood, 2017, as cited by Jordan and Hewitt, 2022). It is also connected to the scientization, technization, and economization of issues addressed by closed circles of experts and organizations (Ylönen et al., 2017). Floyd posits that we can see three situations at play: an issue being nonpoliticized and nonsecuritized, an issue being politicized (i.e., a visible part of party political debates), and an issue being securitized. Further, she argues that when environmental issues are desecuritized, the morally right option is for them to be politicized (by an official political authority). Related to this, Trombetta (2009, p. 589) has argued that desecuritizing the environment "can lead to the depoliticization and marginalization of urgent and serious issues, while leaving the practices associated with security unchallenged." Aradau (2004, p. 393) put forward the idea that, if desecuritization is the opposite of securitization, it is then about the "democratic politics of slow procedures which can be contested." Overall, security studies present differing interpretations of the link between securitization and politicization.

Climate change is an example of the extension of securitization to new domains, where the distinction between securitization and politicization proposed by Floyd is not followed so strictly.[1] However, it is commonplace nowadays for environmental and resource issues to be integrated into governmental security strategies. Berling et al. (2021) identified two kinds of connections between climate change and security: first, these issues are "compared" when policymakers identify and prioritize threats; and, second, climate change may trigger new security concerns.[2] Claire Dupont has studied securitization of climate change in the EU in terms of the "speech acts" performed and has concluded that the collective securitization of climate change has been a success (Dupont, 2019). With this, she means that climate change has become the crucial policy agenda issue it needs to be and does not refer to the tight interpretation of securitization as security politics provided by Floyd. In the context of climate change, securitization is perhaps

---

[1] There are, however, security scholars who want to make a clear distinction between *risks*, that is, the conditions of possibility for harm, and security *threats*, that is, direct causes of harm (Corry, 2011). Floyd (2019, p. 95) states that "threats can be defended against, whereas risks can only be managed."

[2] As a coauthor of the Berling et al. (2021) paper, Ole Wæver has thus acknowledged the reach of securitization beyond traditional national security since the late 1990s.

used more often as a political or policy tool than as part of formal security policy. This is because securitizing the environment is a powerful way to draw attention to otherwise unaddressed issues (Peoples and Vaughan-Williams, 2015). Dupont states that the first attempts to securitize climate change internationally remained at the level of speech acts, because the "audience" of negotiating partners rejected the "securitizing moves" and "extraordinary measures" did not follow the speech acts. Yet, post-2008, several moves resulted in "a new securitized status quo, with climate change firmly embedded in high politics" and measures for mitigating climate change comprising an extraordinary role for the European Council (Dupont, 2019, p. 382).

In the energy context, the volume edited by Kacper Szulecki (2018b) drew a distinction between energy security rhetoric and actual implemented policy measures; extraordinary policies being a rarity and "energy security" rarely being securitized. In the literature on the political economy of energy, extraordinary measures have also been associated with a break from previous political practice (previously such breaks would only have been associated with emergency measures) (Kuzemko, 2014). Heinrich and Szulecki (2018) argued that if extraordinary emergency measures are narrowed down to military interventions, most interesting features of energy securitization would be excluded. However, if such measures refer to removing energy issues from public oversight more interesting analyses emerge. The view of Heinrich and Szulecki about securitization and extraordinary measures connects with the idea of depoliticization. The authors propose three kinds of extraordinary measures in energy policy that would break normal political practices, strengthen the executive powers of selected agencies, or isolate selected decisions and potentially important information from public access: first, breaking norms about "how things are done"; second, shifting power to the agency level; and, third, withholding or limiting information (Heinrich and Szulecki, 2018). Depoliticization is a somewhat less strong process, but it nevertheless removes energy issues from open political debate (Kuzemko, 2014). Interesting questions pertaining to securitization and depoliticization are, for example, who has the power to put such measures in place. In sustainability transition terms, the securitizing actors would normally be sociotechnical regime actors, because niche actors seldom have the power to conduct extraordinary measures until transitions have progressed to a phase whereby a niche is institutionalizing as a result of joint actions between niche and regime actors.

While security often tends to have a negative connotation via a focus on threats, some security scholars (especially from critical security studies) prefer to recognize a positive framing of security; that is, something additional to and not replacing negative security (Hoogensen Gjørv and Bilgic, 2022; Roe, 2008). For example, Ken Booth (2007) emphasizes emancipation, peaceful and positive relations, and freedom from insecurity as positive historical associations with the term security. Further, he argues that the "[l]anguage of securitisation freezes security in a static framework, forever militarised, zero-sum, and confrontational" (Booth, 2007, p. 165). The idea to conceptualize security in positive terms originates from the concept of human security – inspired by concepts of security and peace (Floyd, 2019). According to Cortright (2017), inclusivity, participation, and capacity to ensure security in governance systems can increase the prospects of peace. The conceptualizations of positive and negative security relate to how security is valued: as a buffer against things we wish to avoid and as security discussed no more than necessary (related to negative security), or as a foundation to a good life (positive security). Hoogensen Gjørv (2012) explores the relationship between negative security and positive security, arguing that positive security covers gaps that negative security as a concept misses, and addresses the epistemological foundations used when talking about security. She views negative security as typically connected to so-called traditional security: an epistemology of fear, identifying threats and justifying the use of force based on danger of death. This also relates to the state as the sole actor for security, with little attention paid to multiple voices even within the "state." Conversely, positive security can be perceived via the lens of human security, focusing on individuals but also societal well-being more broadly. Positive security connects to feelings of safety and stability and to the security of expectations, which enable building future capacity (Hoogensen, 2011). However, the concepts of negative and positive security should not be associated with "bad" and "good" but rather with the different approaches of security; for instance, whereas negative security can be focused on the absence of violence, positive security can emphasize the inclusion of social justice (Hoogensen Gjørv and Bilgic, 2022). They can, therefore, be used in a complementary manner.

McSweeney (1999) mentioned the stable character of routines *enabling* creativity in the context of positive security. Hoogensen Gjørv (2012) used this idea more actively to define positive security in terms of *enabling* people and communities – the central foundation of such enabling being trust. She argued that enabling can be conducted either by external actors, for example NGOs, or created within communities. Roe (2008) suggested that positive and negative security can be distinguished by the values that are pursued, where positive values relate to the advancement of justice. Hoogensen Gjørv (2012) proposed a three-step process for understanding both positive and negative security: first, recognizing actors, practices, and the specific context; second, identifying the epistemological foundation of (i.e., assumptions behind) the practices; and, third, looking at the values, such as justice, associated with those practices. More recently, Hoogensen Gjørv and Bilgic (2022, p. 2) stated: "Positive security finds its meanings in its unfolding.

It is about myriad ways of practicing security in multiple daily encounters with the other(s). It is not predetermined or a certainty, but a possibility."

There are interesting connections between positive security and sustainability transitions, starting with the process of unfolding. One can say that the outcomes of sustainability transitions are, likewise, not predetermined. Moreover, the conceptualization of positive security connects to the elements of justice, actors, practices, and assumptions that are core parts of sustainability transitions. For example, on the one hand, enabling people and communities for positive security – as mentioned above – may take place in connection to new sociotechnical niches. Yet, on the other hand, transitions tend to disrupt routines and practices, which may increase feelings of insecurity and reduce positive security. In the empirical parts of this book, I utilize the conceptualizations of positive and negative security and how these two securities are presented via the assumptions, values, and illustrations of practices provided by expert actors at the interface of security and energy transitions.

In this book, I occasionally use the concept of "stability" as linked to security. This refers to the absence of armed or nonarmed violence but can also be more broadly connected with safe and well-managed societies. Cortright et al. (2017, p. 22) state that the "prevention of armed conflict is linked to stable governance structures that have the capacity to deliver public goods to all stakeholders, provide for public participation and accountability, and manage competing claims to power, resources and territory." One way to define stability is as the capacity to maintain state security and the ability to withstand and avoid political and other shocks; in essence a kind of resilience. On an individual level, it has been noted that routines and a regularizing social life establish cognitive stability, which is connected to positive security (Roe, 2008). From that perspective, sustainability transitions, that is, both the shift in practices and the disruption of existing sociotechnical regimes, may invoke temporary cognitive instability, which may explain some part of the resistance to sustainability transitions.

## 2.3 Security in Transitions Research

Security, either as a force molding sustainability transitions or as something that is affected by evolving transitions, has not gained much attention in sociotechnical transition studies. Phil Johnstone and colleagues were the first scholars to begin to make any explicit connection to security in sustainability transitions. They adopted a rather narrow, realist perception of national security as military security, and claimed that conceptualizations of transitions do not consider the "military establishment," and that states, in pursuit of their energy-focused foreign policies, ignore the role of the military (Johnstone and Newell, 2018). Johnstone also used the terms "military industrial complex" and "national security state" to discuss the roles played by incumbent actors with vested interest in the established sociotechnical regime, and their potential strategies to impede the acceleration of niche innovations (Johnstone et al., 2017). In this context, *securitization* was understood as altering policy goals in terms of (traditional) national security, while *masking* was where military security interests are disguised as civil energy policy activities. This links perhaps to two views of securitization indicated by security studies: securitization as open security politics and politicization (when, for example, energy policy is openly linked to security policy goals) and securitization as depoliticization where security and exceptional measures are removed from the "public gaze"; the latter matching the definition of securitization by the Copenhagen School.

More generally, before the early 2020s only a few transition studies discussed security. Regarding pathways for electricity sector transitions, Geert Verbong and Frank Geels referred to geopolitical security and energy security as significant landscape threats (Verbong and Geels, 2010). Geels remarked that the military dimension is a part of fossil fuel alliances composed of incumbent firms and policymakers (Geels, 2014). In the early 2020s, security-related aspects have received somewhat more attention in transition studies – albeit associated in particular with negative security. For example, it has been recognized that the interests of the fossil fuel industry have shaped the perceptions of states, such that some may resort to war in order to secure critical resources for their sociotechnical energy regimes (Ford and Newell, 2021). Transition concepts have also been used to study the increase of renewable energy in conflict and postconflict regions (Chaar et al., 2020; Fischhendler et al., 2021). Kester et al. (2020) applied a critical security studies lens to the study of mobility transitions and referred to negative and positive security. They pointed out, for example, how visions or expectations based on negative security or securitization can hinder niche development. A study on food system transitions noted that efforts to transition diets should consider existing injustices in food security "to reduce the overall vulnerability of those groups who are prone to transition-inflicted harms" (Kaljonen et al., 2021, p. 481).

An interesting development is recent research on deep transitions. Deep transitions have been described as the transformative changes of multiple sociotechnical systems in a similar direction. Historically, this directionality has comprised, for example, reliance on fossil fuels and global value chains alongside resource and energy intensity (Schot and Kanger, 2018). In this context, Johnstone and McLeish (2022) have explored the relationship between the world wars and multisystem sociotechnical change. They showed, for example, that World War II helped stabilize and internationalize the supply and use of oil as a key energy source. Further, energy, food, and transport systems were coordinated toward a similar direction to win the war, resulting in a "consolidation of meta-rules" (Johnstone and McLeish, 2022, p. 12). Also, others have noted the role that militaries have historically had in building new infrastructure systems, such as railroads (Van der Vleuten, 2019). While these new infrastructure developments had elements that linked to positive security, they were largely supporting negative security efforts: "[M]ilitary system builders captured and appropriated the same mobility transition that ought to bring peace, progress, and liberty, only to develop unprecedented warfare capabilities and scales of violence" by entangling the ongoing transport transition with the transformation of the military system (Van der Vleuten, 2019, p. 30). Therefore, we must also exercise some caution with ongoing transitions and not automatically assume they only have benign connotations.

The impacts of the world wars on sociotechnical transitions were long-lasting. In addition to advances in technology and infrastructure, the wars altered the wider cultural context via memories and expectations around the potential for another war, resulting in an "upward" effect on the sociotechnical landscape (Johnstone and McLeish, 2022). While the kind of demand for sociotechnical change resulting from the two world wars is unlikely to be seen as a result of the war in Ukraine, the latter is, nevertheless, creating openings for new niche innovation (e.g., small modular nuclear reactors), the wider expansion of existing niches (e.g., wind power), and the destabilization of dominant technological systems and institutions (e.g., oil) (with cascading effects on global energy and food systems). It is likely to lead to significant changes in European energy regimes.

The rapidly expanding literature that connects justice to sustainability transitions may also be important from the perspective of security, particularly positive security. Jenkins et al. (2018) introduced the concept of energy justice to sustainability transition studies. They argued that calls for transitions need to include concerns for a fair distribution of infrastructure and services, equal access to decision-making, and promoting participation of marginalized groups. To describe justice, they referred to its three tenets: distributive, recognitive, and procedural.

Distributive justice refers to the equal distribution of monetary and nonmonetary costs and benefits of a transition or a policy action. Recognitive justice is focused on how those in more vulnerable or marginal positions in society are impacted or taken into account in decision-making; and procedural justice pays attention to participation opportunities, and the fairness and transparency of policymaking processes (Jenkins et al., 2018).

Addressing such concerns of justice is likely to contribute to positive security in transitions. This can happen by enabling people and communities, as described by Hoogensen Gjørv (2012). Yet tensions and resistance to transitions may arise as a result of experiences or perceptions of injustice or a lack of democratic decision-making (Healy and Barry, 2017; Jenkins et al., 2018). For example, in

Australian coal communities anxiety over employment has led to social and political resistance to phasing-out coal, and to hostility toward the just transition concept itself (MacNeil and Beauman, 2022). Other studies have pointed out connections between right-wing populist politics and resistance to transitions (Abraham, 2019; Żuk and Szulecki, 2020). Further, Abraham (2019) has argued for making just transitions a populist concept due to its ineffectiveness to shield against populism. Therefore, the ways in which justice and injustice are perceived (rather than realized) influences how sustainability transitions unfold – via the absence or presence of tensions that may escalate into conflicts. Nevertheless, despite perceptions, transitions may also result in increased or decreased justice in effect; for example, by advancing solutions that promote peace and stability or by heightening inequalities between different groups of people (Kivimaa et al., 2022).

Geopolitical risks caused both by climate change impacts and by the efforts to mitigate climate change concern questions of justice at different levels. Initially local conflicts, spurred by either of the abovementioned, may cascade into security risks in larger regions or internationally (Carter et al., 2021). Yet research indicates that climate change is likely to induce larger geopolitical risks than its mitigation. For example, energy transitions have been envisaged to reduce the number of large conflicts between countries and regions (Vakulchuk et al., 2020). The research on the geopolitics of the energy transition is addressed in Chapter 3.

Finally, it is pertinent to note that, while transition studies have paid relatively little attention to security connections, it draws from innovation studies that have originated from science and technology studies and the history of the world wars. Science, technology, and innovation (STI) policy was created in the aftermath of World War II. After the war, concerns about future economic recovery initiated STI policies that aimed for growth, mass production, and consumption; these policies expanded the role of the state in advancing scientific research, with the idea of also helping to maintain peace (Schot and Steinmueller, 2018). In the US, postwar STI policy explicitly stated a contribution to national security to be one of the tasks of government STI policy (Lundvall and Borrás, 2005). Later, the Cold War spurred on defence-related research and development (R&D) and contributed to the development of national innovation systems, while the pursuit of economic growth gradually became the dominating goal (Mowery, 2012). The economic growth agenda has subsequently led to multiple severe environmental problems, including climate change, unsustainable levels of resource use, pollution, and overexploitation of natural environments (Kivimaa, 2022b), and hence to the development of the field of sustainability transitions.

# 3
# Energy Security and Geopolitics of Energy Transition

This chapter's objective is to situate this book within current knowledge and past developments in energy security and geopolitics research. Previous accounts of energy security and geopolitics have typically been limited to the energy perspective and have failed to delve into many of the broader questions of security – as outlined in Chapter 2. This chapter starts with a summary of the conceptualization and history of energy security research, which is largely focused on the differing definitions of energy security. The chapter then moves onto the more recent and rapidly increasing literature on the geopolitics of the energy transition and, in particular, the geopolitics of renewable energy. Much of this literature is based on descriptions of potential future trajectories of how the geopolitics of renewable energy or energy transition will unfold, rather than empirical research. However, the literatures on energy security and geopolitics of renewable energy are an important context into which the empirical analysis documented in this book is placed. The chapter ends with a brief account of energy security in Europe.

## 3.1 Conceptualization and History of Energy Security Research

Energy security research can be divided into conceptual and empirical studies. In this subsection, I briefly review the conceptual development of energy security.

Energy security began to attract political interest as a result of the world wars, as the oil infrastructure was expanded to support oil supply to the military (Johnstone and McLeish, 2022). The use of oil grew in the aftermath of the world wars, leading to an increased reliance on oil in society (Chester, 2010). Academic research on energy security at the time was rare, but some early writings appeared in the 1960s (Lubell, 1961). However, following the 1970s oil crises, energy security emerged as an issue on many states' political agendas. In some states, such as Finland, energy policy became then distinguished as a specific policy domain for the first time.

In 1976, Willrich provided alternative definitions for the concept of energy security, such as: "the guarantee of sufficient energy supplies to permit a country to function during war"; "the assurance of adequate energy supplies to maintain the national economy at a normal level"; and "the assurance of sufficient energy supplies to permit the national economy to function in a politically acceptable manner" (Willrich, 1976, p. 747). All of these approaches took "the nation," rather than citizens, to be the entity for whom energy was to be secured. Willrich also stated that energy security was, in all approaches, closely linked with economic security. However, even in the 1970s, he acknowledged the importance of addressing the environmental impacts of energy production alongside energy security, as part of energy governance. He discussed accidents, land-use issues, air and water pollution, radioactive waste, climate change (using the term "thermal limit"), and energy conservation. Many issues and mechanisms that are today discussed in conjunction with energy security were mentioned by Willrich. Alongside the environment, these included, for instance, self-sufficiency, stockpiling, and assuring foreign supplies. However, although some academic research emerged in the aftermath of the 1970s (Yergin, 1988), scholarly attention to energy security decreased because of stabilizing oil prices in the 1980s and 1990s (Cherp and Jewell, 2014).

The research field began expanding after 2000, following rising global energy demand, concern over gas supply, and decarbonization pursuits. In particular, the period from 2006 to 2010 saw significant developments in energy security studies (Azzuni and Breyer, 2018). This research was in stark contrast with earlier research, both because it was no longer focused solely on oil and because it began to provide multiple interpretations of energy security in diverse contexts.

In a similar way to the concept of security, there have been multiple definitions of the energy security concept. Drawing from broader security studies, Cherp and Jewell (2014, p. 417, emphasis in original) defined energy security as *"low vulnerability of vital energy systems."* Jewell and Brutschin (2021) later specified this as the absence of threats and the capabilities of states and system operations to respond to threats. Cherp and Jewell (2014) proposed three key questions for energy security: Security for whom (e.g., households, industry, or states)? Security for what values (e.g., political, economic, or social)? And security from what threats (e.g., natural weather events, terrorist and military attacks, or technical disruptions)? They also talked about vulnerability in terms of the diverse nature and origin of risks. Such risks have been divided by Winzer (2012) into technical, human, and natural risk sources. Winzer describes technical risks as infrastructure interdependencies, mechanical/thermal failures, and emissions. Human risks are linked, for example, to geopolitical instability, political instability, and terrorism. Natural risk sources not only refer to natural disasters but also to resource intermittency and depletion. Winzer argued that different risks are not of similar magnitude, but have differing scopes, speeds, durations, and severities of impact. Like the sustainability transitions concept of "landscape," he makes a distinction between "shocks" as short-term disruptions and more longer-term "stresses" (Winzer, 2012).

In the 2000s, extensively cited work introduced the four As of energy security: availability, accessibility, affordability, and acceptability (Kruyt et al., 2009). Cherp and Jewell (2014), however, criticized the perceived importance of the four As and argued that dimensions such as acceptability should not be included in the definition of energy security as they confuse its interpretation. Yet others have elaborated multiple dimensions of this concept. For example, Sovacool and Mukherjee (2011) proposed twenty components of energy security under five dimensions: availability, affordability, technology development and efficiency, environmental and social sustainability, and regulation and governance. Some of these components were more directly related to the supply of energy, such as availability and dependency, while others were connected instead to other societal objectives, such as water, land use, pollution, and greenhouse gas emissions. Systematically reviewing the literature on energy security, Azzuni and Breyer (2018) identified fourteen dimensions and parameters for energy security: availability, diversity, cost, technology and efficiency, location, timeframe, resilience, environment, health, culture, literacy, employment, policy, and the military.

In the 2010s, the energy security literature begun to more extensively include the question of climate change and the need to decarbonize. These connections, however, have mostly been made in terms of empirical studies (Knox-Hayes et al., 2013; Rogers-Hayden et al., 2011; Strambo et al., 2015; Toke and Vezirgiannidou, 2013). The studies did not propose a principled priority of decarbonization over other energy security dimensions and provided few conceptual insights into energy security. Another review of energy security studies, however, showed some change in the emphasis of the concept over time; while energy availability has constantly remained as a key element, energy prices, efficiency, and the environment have increasingly been discussed in association with energy security (Ang et al., 2015).

Chester (2010, p. 887) has accurately remarked that "the concept of energy security is inherently slippery because it is polysemic in nature, capable of holding multiple dimensions and taking on different specificities depending on the country (or continent), timeframe or energy source to which it is applied." He noted that energy security is typically used to refer to unhindered and uninterrupted access to energy sources, a diversity of sources, nondependency on a particular geographical region for energy sources, abundant energy sources, some form of energy self-sufficiency, and/or an energy supply that can withstand external shocks. Overall, energy security has been described as a context-specific political phenomenon (Knox-Hayes et al., 2013; Szulecki, 2018a). For example, Winzer (2012) pointed to political differences between states, where energy security has been associated with energy independence, energy diversity, reliability of supply, or protecting the poor against energy price volatility. Variation in understanding energy security can perhaps be explained by differences in how actors value different parameters, such as the resource sufficiency or import dependency of a country or market-based solutions versus state involvement (Månsson et al., 2014). Most discussions on energy security have tended to debate it from the perspective of states, and the academic literature has frequently ignored energy poverty as a security question. The individual household perspective in terms of energy security did, however, become more visible in discussions around the European energy crisis in late summer and early fall 2022.

These more diverse dimensions to energy security create a challenge for achieving a coherent energy policy, while the simpler framing of energy security – the availability of adequate energy at an acceptable price – matches the purely economic understanding of energy policy and largely ignores the geopolitical dimension (Dyer, 2016). More in-depth discussion on the concept of energy security can be found in a book edited by Szulecki (2018a), which also deliberates the difference between inductive and deductive ways to define the concept.

In summary, the concept of energy security has broadened to new dimensions over time. Doing so, it has become analytically less meaningful or "slippery" as some have described it. Thus, the ways in which energy security has in practice been applied – and what dimensions are emphasized – are contingent on the values guiding policymaking in given contexts. In this book, I adopt the relatively simple definition of energy security by Cherp and Jewell (2014, p. 417, emphasis in original) – "*low vulnerability of vital energy systems*" – but acknowledge its problems in that it does not extend "security" beyond the operational security of the energy system itself. Therefore, drawing on the multitude of dimensions proposed for energy security, I argue that one can in essence talk about "internal energy security," which connects to the definition by Cherp and Jewell and the secure operation of the energy system itself (see Figure 3.1). In addition, this internal energy security can be distinguished from "external energy security," which addresses the broader security implications of the energy system. These include the effects of energy installations and infrastructure on the environment, that is, environmental security, and on human health and well-being, that is, human security. For example, nuclear radiation leakages (accidental or purposefully instigated) can cause both human health and the environment to deteriorate. The 2022 Nord Stream pipeline gas leaks into the waters of Sweden and Denmark show that fossil fuel infrastructure can be used to harm the environment or humans. In addition, renewable energy sources, such as hydropower or wind power, can have negative environmental security implications.

## 3.2 Geopolitics of Renewables

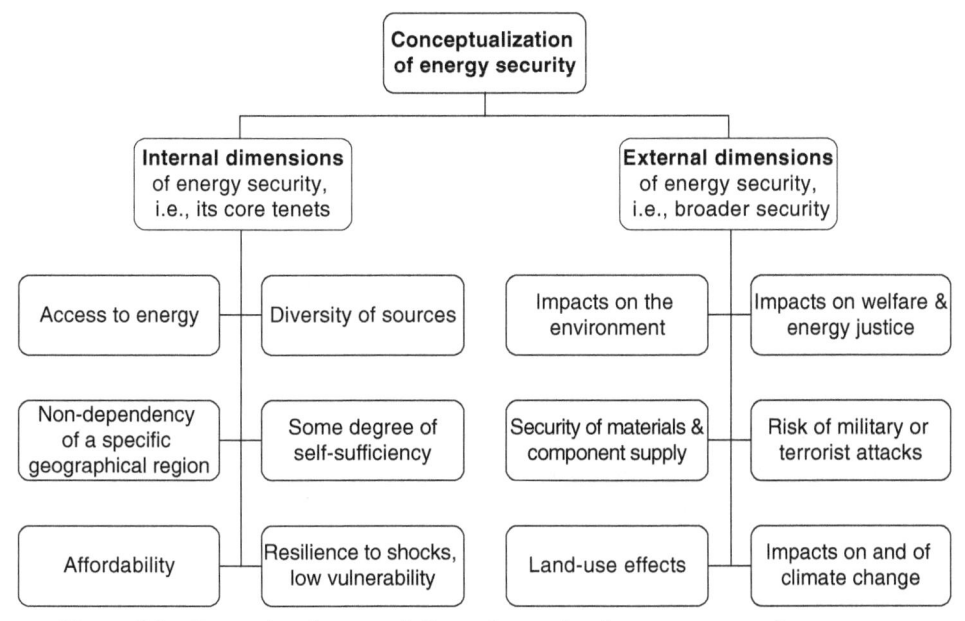

Figure 3.1 Internal and external dimensions related to energy security.

In this book, I use the term energy security to refer to internal energy security, while what falls under the external categories can be addressed in terms of broader security implications. Regarding the former, I propose it be split into several sub-areas. Drawing from Sovacool and Mukherjee (2011), these include, for example, secure supply of required fuels, minerals, and technical components (typically provided to a large degree via international trade); secure supply of electricity; reliability of production against technical faults and weather-induced disruptions; diversification of energy sources; sufficient domestic energy supply; and stockpiles of fuels or electricity storage for emergency situations. Issues such as military or terrorist attacks on energy infrastructure or the climate and environmental security of energy installations are thus not covered under energy security but under the broader conceptualization of security (see Chapter 2).

### 3.2 Geopolitics of Renewables

Security as a concept is connected to geopolitics, which can be seen as one dimension related to security. The geopolitics of energy has explored the global energy regime and the ways in which energy relations among producer, transit, and consumer countries advance and impact international relations (Criekemans, 2018). Classical geopolitics defines geopolitics as the effect of geographical factors (e.g., a country's size, position, or resources) on international relations and the power of states (Kelly, 2006; Overland, 2019). It "emphasises the international role of the state in energy in terms of securing supply, engaging in strategic alliances, and exercising military power, with access to energy resources seen as a zero sum game" (Kuzemko et al., 2016, p. 9).[1] An example of a classical assumption is that abundant resources are seen to correlate with geopolitical influence and unequal resource access can spur international conflicts (Pflugmann and De Blasio, 2020). For instance, vast fossil fuel resources have granted Norway greater geopolitical power than a country of this size would otherwise have (see Chapter 7).

Critical geopolitics, however, raises doubts about the pregiven role of geographical factors in international relations and seeks to reveal how geographical beliefs are used in global politics (Kuus, 2017). This strand of the geopolitics literature questions dominant power structures and knowledge (Tuathail, 1999). Critical geopolitics reveals that geopolitical competition over energy resources is socially constructed and at least partly imaginary (Blondeel et al., 2021). This means that not all similarly resource-rich countries are equally powerful. What we can draw from this to apply to the energy transitions context is that geopolitical beliefs have both overt and concealed influence in energy political decisions (Overland, 2019; Vakulchuk et al., 2020), which pertain to how transitions are advanced or hindered by government politics. Geopolitical beliefs and assumptions can either enhance the role of security in energy policymaking, as in Estonia, or emphasize cross-border economic relations, as in Finland prior to 2022 (see Chapters 5 and 6). Moreover, they can influence how the benefits and drawbacks of alternative energy systems based on renewable energy are perceived in security terms.

During the 2010s and early 2020s, much research was undertaken in connection to the geopolitics of renewable energy, and, later, of hydrogen, because of an increasing ambition to mitigate climate change. This was preceded by established research on the geopolitics of hydrocarbons, especially oil (Victor et al., 2006; Yergin, 2009). Oil had become a strategic resource during World Wars I and II and later stabilized as a key factor of the global energy regime (Johnstone and McLeish, 2022). This development was further supported by the diffusion of private cars and growing car ownership, making Western countries dependent on oil imports and oil a key issue within economic stability (Overland, 2019).

Later, natural gas became a hot topic in the geopolitics of energy literature. One stream of this literature addressed EU–Russia energy relations, especially after the 2006 and 2009 gas crises in Ukraine caused by Russia (Sharples, 2016; Siddi, 2018). The literature on the geopolitics of hydrocarbons has typically not considered decarbonization (Van de Graaf, 2018).

---

[1] Geopolitics of energy connects to the broader research area of the international political economy of energy, which addresses the governance of energy issues from the perspective of altering the balance between state and market activities (Kuzemko et al., 2016).

The literature on the geopolitics of renewables departs from an argument that the geographic plentifulness of renewable energy sources will influence cross-border energy flows. These flows have traditionally been based on hydrocarbon resources, that is, the international supply of oil, gas, and coal. Hence, the expansion of renewable energy changes the ways in which states interact with regard to energy issues, and also presents new challenges for energy trade and energy security (Scholten and Bosman, 2016). In the mid-2010s, issues such as access to technology, power lines, rare earth minerals, patents, storage areas, and dispatch methods were used to formulate the new geopolitics of low-carbon energy sources. Paltsev (2016) also noted that geopolitical power relations are influenced by the timing and stringency of climate policies. Besides decarbonization developments, the geopolitics of energy relations is affected by strengthening Asian economies contributing to globally rising energy and minerals demands that will potentially result in energy scarcities (Criekemans, 2018).

The methods used in the research on the geopolitics of renewables range from hypothetical cases or thought experiments (Scholten and Bosman, 2016) to document analyses (Koch and Tynkkynen, 2021). In rare cases, remote sensing has been applied as the method of choice (Fischhendler et al., 2021). Approaches using critical geopolitics seem rarer (e.g., Koch and Tynkkynen, 2021; Overland, 2019). Yet, many even widely cited pieces from this literature are speculative perspective articles with relatively little conceptual or empirically supported insights. This differs greatly from approaches in sustainability transitions research, which typically require conceptually informed new empirical research. Indeed, international relations scholars themselves have noted that no specific theory has been formulated around the geopolitics of renewable energy (Vakulchuk et al., 2020).

Broadly, scholars looking at the geopolitics of renewables see many *positive outcomes* from the expansion of renewable energy. For example, Scholten et al. (2020) describe a positive disruption that brings forth new challenges for energy security. They point out the benefits of renewable energy resources compared to fossil fuels – continuous and variable as opposed to geographically concentrated and exhaustible, allowing decentralized generation – as well as drawbacks, such as the need for relatively large amounts of critical materials and metals and the fact distribution is mostly via electricity networks. Kuzemko et al. (2016, p. 162) state that "[f]rom a climate perspective, the shift to a low-carbon energy pathway will result in far greater energy security." On the other side of this discussion are the geopolitical implications of the transition on the hydrocarbon sector (Blondeel et al., 2021; Van de Graaf, 2018), which may reduce global stability, at least in the medium term. The literature effectively, therefore, highlights two phases: the *transition phase* and its effects on security, as well as the *later phase* when new systems have formed and stabilized. The transition phase is expected to destabilize global security and cause tensions between winners and losers (Blondeel et al., 2021; Vakulchuk et al., 2020). During the transition phase, established trade relations are likely to break down and new partnerships form (Scholten et al., 2020). The material flows make China an important actor in the new geopolitics of energy. The new stabilized system, in turn, is likely to benefit from the expected positive geopolitical outcomes of renewable energy more fully.

While most of the geopolitics of renewables literature addresses global dynamics, selected studies have focused on patterns of conflict and cooperation in specific geographical contexts. For example, pertaining to the region of Israel and Palestine, Fischhendler et al. (2021) have shown how renewable energy can also diffuse rapidly in conditions of armed conflict. More specifically, they observed that the Gaza Strip became a regional leader in solar energy, but this has required that Israeli policymakers not consider solar energy technology to be a security threat.

Drawing from an idea in a coauthored paper (Kivimaa et al., 2022), I now focus on three (sociotechnical) components via which the geopolitical implications of energy transitions can be addressed: technology, actors, and institutions. I must note, however, that these categories are interconnected and are here discussed separately purely for improved clarity.

### 3.2.1 Technology

Technological change plays a major role in the geopolitics of energy (Criekemans, 2018). Further, it is well acknowledged that the energy transition based on renewables will bring forth technical and system challenges due to its intermittent nature, affecting "internal" energy-system security. This means, for example, that certain renewable energy sources, solar and wind, cannot be produced to meet demand at any given point in time but depend on the weather (Scholten and Bosman, 2016). Scholten et al. (2020) note that smart technologies, demand-side management, and spatial distribution are vital for balancing the electricity system. On the positive side, when connected to decentralization of production, renewable energy systems are expected to experience a smaller magnitude of harm from disruptions and affect fewer people (Groves et al., 2021). New technical configurations are needed to create new reliable energy systems (Child et al., 2019).

Electricity is the main carrier for many renewable energy technologies, such as solar, wind, and hydropower. This "implies a physically integrated infrastructure that connects producer and consumer countries through a single interconnected grid" (Scholten and Bosman, 2016, p. 227). It also requires electrification to have an increasing role in this transition; this comes with its own share of geopolitical consequences as well as technical security considerations. On the technical side, electricity is not as flexible as solid fuels because demand and supply must meet at any given point; this also gives rise to a rather complex organization of spot and futures markets. Any disturbance to the system may, in the worst case, affect the whole network. Some means for storing electricity exist, such as pumped hydropower storage and batteries, but much technological development is still required (Scholten and Bosman, 2016). It has been argued that the electricity trade based on renewables creates more symmetrical connections between countries, whereby several countries produce renewable electricity but exchange with neighboring countries to balance their grid (Overland, 2019). Even large international "supergrids" have been part of the discussions and plans. These supergrids may improve technical energy security by reducing the supply-related disruptions associated with long-distance shipping of hydrocarbons (Scholten, 2018) and the magnitude of country-specific backup reserves (Blondeel et al., 2021; Scholten et al., 2020). On the other hand, new system vulnerabilities are expected, such as the growing potential of and surface area for cyberattacks (Cornell, 2019).

In the geopolitics of renewables literature, resource-based dependency on critical materials and renewable energy technology has increasingly been discussed. More minerals and metals are needed when renewables-based systems expand. Some critical materials are described as "rare earths" and, even within rare earths, and indeed all critical materials, some are rarer or more valuable than others. Renewable energy technologies need a range of materials, such as cobalt, lithium, aluminum, dysprosium, and neodymium. When demand expands, the cost of the materials and elements is expected to increase (Paltsev, 2016). Lithium has already proved critical and, thus, replacements are increasingly being sought (Greim et al., 2020). Some scholars argue that we do not yet know the scale and scope of the security challenges brought by critical materials (Lee et al., 2020; Scholten et al., 2020). Nevertheless, international actors, such as the International Energy Agency (IEA) and the EU, are increasingly investigating the geopolitical implications and security of supply around critical materials (EC, 2020; IEA, 2021).

The "resource curse" is mentioned as one issue in the energy geopolitics literature. This refers to an illogically slow growth of resource use in resource-rich countries combined with slow economic growth, high income and gender inequalities, a low level of democracy, and negative social, environmental, and economic impacts (Hancock and Sovacool, 2018; Leonard et al., 2022). In the context of renewables, the resource curse has been mentioned in relation to critical materials, metals, and metalloids (Månberger and Johansson, 2019) as well as hydropower (Hancock and Sovacool, 2018). Critical materials are unevenly distributed among countries, but less unevenly than hydrocarbons. These minerals and metals can be found around the world, but countries have dissimilar opportunities to extract them, leading to a range of security and geopolitical consequences (Månberger and Johansson, 2019). With respect to renewable energy sources, such as wind and solar power, this means that even in countries where the local climate and weather conditions are favorable for their expansion, they do not form significant energy sources unless many politicians support their development. Lederer (2022) has described this as politics trumping geography. This connects to the next theme: actors.

### 3.2.2 Actors

The geopolitics of renewables literature emphasizes the actor dimension via relations between states as actors. Many scholars argue that the expansion of renewable energy changes interstate power relations (Criekemans, 2018; Johansson, 2013; Overland, 2019; Scholten and Bosman, 2016). In addition, some of the literature also addresses intrastate tensions and conflicts.

The energy transition means moving toward less oligopolistic markets and more symmetrical energy relations, as most countries can produce some form(s) of renewable energy. This reduces geopolitical risks for those states that have previously been dependent on hydrocarbon supply from others (Blondeel et al., 2021). In the transition phase, states can make a decision between (inexpensive) imported energy and more secure domestic renewable energy reserves. Scholten et al. (2020) argue that the new global energy system may dilute differences between previous import and export countries, creating a world of "prosumer" countries. While the potential to possess renewable energy exists for all countries, some countries may be "richer" in terms of annual solar radiation or potential areas for wind power. Countries with abundant hydropower reserves, such as Norway, have the benefit of balancing capacity that other countries lack, linking to the technical aspects described in Section 3.2.1.

Countries dependent on exporting fossil fuels may became destabilized or more unstable than they already are (Kuzemko et al., 2016) and have been envisaged as those losing most from the energy transition (Vakulchuk et al., 2020). Van de Graaf (2018) discusses three strategies such petrostates may follow when reacting to the energy transition. The first is that they will "race" to sell oil. This means more oil is extracted from the ground as long as the present demand continues, potentially leading to "price wars" between oil producers. The second strategy is to preserve profits from oil for the future by curtailing production. This strategy opposes the first one and is likely to require agreements among oil-producing countries for specific production quotas. The third strategy concerns domestic economic reform, which means broader transformation in how (instead of oil) revenue is generated for the country in question. If global energy transition succeeds, this will be a necessary strategy for all hydrocarbon states.

Scholten et al. (2020) deliberate whether the decentralization of production via renewables and electrification will lead to an overall reduction in international trade or change the shape of trade from fuels to renewable energy production technologies and energy services. They argue that hydrocarbon-related tensions will dilute, shifting investments into renewables; at the same time, however, globally increasing energy demand may reduce the positive effects of renewables and prevent them becoming a strategic factor because fossil fuels are still needed to meet rising demand. Some have forecast that the number of large conflicts will decrease (Vakulchuk et al., 2020). However, Blondeel et al. (2021) highlight that decentralization does not automatically lead to decreased tensions and, for example, energy self-sufficiency may provide less incentive for countries to avoid conflicts because they are less dependent on each other in energy terms.

At the turn of 2010s, rare earth materials emerged as a significant issue in Asian security policy, especially following China's embargo on the supply of rare earths to Japan in 2010 (Wilson, 2018). As the crisis abated quickly, rare earth materials did not receive similar attention in Europe until about a decade later. In academic research, the issue of critical materials in relation to the energy transition was first identified in the early 2010s (e.g., Smith Stegen, 2015). Yet, in the small countries that are the focus of this book, the issue was rarely addressed in public policy documents until around 2022–2023. Research speculates that, despite critical material deposits being quite widely spread between different countries and continents, China's dominance in producing these materials (as the owner of extraction and processing facilities in China and elsewhere and having control over the supply chains) leads to the risk of geopolitical conflicts. Critical material deposits may be used as a "resource weapon," whereby a producer country ends or limits the sale of materials to another country, or, in the worse cases, may spark "resource wars"; that is, armed conflicts over the control of critical materials (Wilson, 2018). Resource scarcities may also lead to internal conflicts within states, where dissidents in weak states use revenues from rare earths to fund their illegal or violent activities (Månberger and Johansson, 2019).

Linking to the technological aspect, circular economy and alternative materials are being developed to reduce requirements for materials. Some scholars argue that the risk of geopolitical competition over critical materials for renewable energy is limited (Overland, 2019). Yet the current need for materials from the energy sector is still too large, in combination with the demand from manufacturers of other sectors' digital technologies, to ensure critical materials are not a security-of-supply issue. One barrier is created by the renewables industries themselves, who regard alternative material solutions as socio-technical niches (Koese et al., 2022). In the meantime, China is a major player, with circa 90 percent share of the market for rare earth minerals and the most integrated supply chains, despite the fact the geographical area of the country itself holds only 39 percent of the world's rare earth reserves (Smith Stegen, 2015). Nonetheless, China is a mineral-rich country that has also acquired mines elsewhere and has set conditions ensuring foreign companies can only use Chinese minerals in production located in China and in collaboration with Chinese companies (Criekemans, 2018; Freeman, 2018). This is a rapidly developing area and the situation has likely further developed since the writing of this book.

Electrification is deeply connected to a renewables-based transition because electricity is the energy carrier for many renewables; it is addressed here from the actor perspective. Scholars have speculated that a regionalization in energy relations may occur, whereby global energy networks based on hydrocarbons change to regional supergrids (Kuzemko et al., 2016). Scholten et al. (2020, p. 3) describe this as leading to new kinds of trade routes and partners, a potentially "fragmented multipolar electric world," providing the example of the Baltic States' desynchronization from Russia even before the 2022 crisis (see Chapter 5). Whereas the geopolitics literature has described regional "grid communities" as improving security (Pflugmann and De Blasio, 2020; Scholten et al., 2020), the 2022 developments also show that there are security risks involved. For instance, the deliberations of Norway – a key producer in the Nordic electricity trading system Nord Pool – to limit electricity transmission to other countries in order to keep their own prices lower somewhat hampered the stable energy relations between the Nordic countries in 2022, besides being a violation of electricity market rules. This links to a Scholten et al. (2020, p. 4) article that emphasized the "reliability of energy partners and the political economic capability to enforce agreements," because the countries forming "grid communities" differ in terms of their economic wealth and political power. Overland (2019) stated that electricity is not well suited as a geopolitical instrument of power. Consequently, any deliberations of the geopolitics literature will be tested in real-life crisis situations.

From the perspective of the local scale, energy transitions may be broadly beneficial. Decentralized modes of renewable energy can facilitate local empowerment (Scholten et al., 2020) and thereby create positive security (see Chapter 2). Power is seen to become more diffuse among states and people within states (Scholten and Bosman, 2016). Criekemans (2018) described potential for a societal revolution, where local and regional groups can organize independently from the state. However, in the transition phase, tensions and resistance to the transition – especially from those who are losing out but also those spreading populism – may lead to civil unrest and separatism (Scholten et al., 2020). Risks of social conflicts have been described, for example, via reduced demand from Europe for Algerian hydrocarbons (Desmidt, 2021), or via right-wing populist parties opposing decarbonization (Vihma et al., 2021; Żuk and Szulecki, 2020). Indeed, energy issues are prone to tensions. Fuel price-related riots have occurred in over forty countries since 2005, with substantial consequences for ordinary people due to their disruptiveness and the violence involved; subsequent policy dialogue is also thus made more difficult (McCulloch et al., 2022). Tensions and conflicts are also likely around land use, which faces multiple pressures. Alongside renewable energy requiring large land areas, for instance, in Lapland – the home of Europe's Indigenous Sámi people – such land-use pressure occurs together with pressures from the effects of climate change, tourism, and mining. These have a combined negative effect on the cultural livelihoods, such as managing grazing for reindeer herds, and the natural environment. Similar examples of several coinciding land-use pressures can be found elsewhere too.

### 3.2.3 Institutions

Institutions include public policy and regulatory structures, formal market structures, and informal structures that have over time formed around sociotechnical systems, comprising the "rules" of the regime (Kivimaa et al., 2022). The geopolitics of renewables literature addresses institutional features less than actors, but some insights can also be drawn here. For example, Scholten et al. (2020, p. 3) state that "we are already witnessing a process of creative destruction in global energy markets." This links to the idea of disruption in sustainability transitions, which implies that not only technologies but markets are disrupted too (Johnstone et al., 2020; Kivimaa et al., 2021). Therefore, the transitional phase and the new energy system are likely to be substantially different from the perspective of international market institutions. This means that governments need to adjust to energy transitions by rethinking national tax systems and energy market designs (Scholten et al., 2020). The 2022 energy crisis in Europe showed that countries were largely unprepared for this. During the transition phase, Scholten et al. describe, for example, the need to create shorter-term intraday markets for renewable electricity to handle intermittency, which is likely to impact market design, regulatory structure, and energy policy practices.

Another aspect is how energy transitions shape foreign and security policy institutions and their interplay with energy policy. The energy objectives of foreign policy have traditionally focused on creating and maintaining alliances that secure (fossil) energy flows to import-dependent countries and promote (fossil) energy exports for hydrocarbon-rich countries. Technological advances in renewable energy and the growing importance of climate policy have created a new strategic objective for energy foreign policy: to "exert influence and reap economic benefits in an emerging low-carbon energy landscape" (Quitzow and Thielges, 2022, p. 599). Energy transition has in essence begun to reorient the focus of energy diplomacy as states' energy relations alter (Griffiths, 2019). For instance, Germany has specifically employed energy partnerships as a form of soft power in foreign relations to gain support for the *Energiewende*, the German energy transition (Quitzow and Thielges, 2022). The institutional shift means also that energy security may become defined more in terms of distribution than energy sources, and via cooperation (Scholten et al. 2020). It has already been observed how reduced hydrocarbon dependencies have somewhat shifted the orientation of foreign policy away from energy security (Mata Pérez et al., 2019), although it did, to some extent, shift back again as a result of the events of 2022. The literature also reports examples where renewable energy has facilitated changes in foreign policy institutions to enable improved collaboration and peacebuilding in conflict areas (e.g., Huda, 2020).

The energy transition also affects international and multilateral organizations. For instance, there have been institutional changes in the impact and membership of the Organization of Petroleum Exporting Countries (OPEC) and the IEA (Bazilian et al., 2017). The IEA was set up in 1974 to promote security of supply for oil and oil markets but has changed its mission to "shape a secure and sustainable energy future for all" (IEA 2024) The International Renewable Energy Agency (IRENA) was established in 2019 and has published influential studies, for instance, on the geopolitics of renewables. However, the OPEC states have not yet undergone substantial energy transition developments despite opportunities to exploit renewable energy (Onifade et al., 2021).

Sanctions have a long history and have been used to create pressure on the target countries by impacting economic relations between states, typically including "restricting exports and imports, freezing assets, and depriving states of financial and economic aid" (Fischhendler et al., 2017, p. 62). The energy sector has been an important target area for placing sanctions. Fischhendler et al. (2017), however, found in their review of the use of energy sanctions that electricity was sanctioned in so few instances that they were unable to analyze this. Thus, it is unclear how the energy transition will influence the use of energy sanctions in the future.

## 3.3 Energy Security in Europe

Before 2006, the energy policies of the EU and its member states had become more and more shaped by market forces and a parting of energy issues from politics, and less influenced by the energy security concern that had emerged in the 1970s (Umbach, 2010). Market liberalization was gradually advanced via the 1998 directive on energy market liberalization and two energy packages in 2003 and 2009, while the energy sector was still organized around national markets (Kuzemko et al., 2016). However, energy security began to attract more attention in Europe as a policy goal in the aftermath of the two natural gas disputes between Russia and Ukraine, in 2006 and 2009. This coincided with the enlargement of the

EU to include Eastern European member states during 2004–2007, which resulted in a larger variance of energy systems (Szulecki, 2018b), and an increase in concern by Eastern European states, particularly the Baltic countries, relating to Europe's dependence on Russian energy sources (Wrange and Bengtsson, 2019). Despite this renewed interest in energy security, the twenty-seven EU member states failed to formulate a coherent strategy for European energy security and energy foreign policy – arguably explained by a lack of political solidarity after the first Russia–Ukraine energy dispute (Umbach, 2010). It was only in 2010 that the EU began to demand member states maintain strategic stocks of natural gas (Kuzemko et al., 2016).

The North Atlantic Treaty Organization (NATO) also began to seriously consider energy security in 2010. It included energy security among its strategic concepts, particularly driven by those Central and Eastern European NATO members that had substantial dependency on Russian energy imports; their gas and energy system development dating back to Soviet times (Bocse, 2020). This was also connected to the EU's energy security debates via the engagement of Central and Eastern European countries and information exchange between NATO and the EU.

A stronger drive for European energy security did not occur until after Russian annexation of Crimea in March 2014 and the lasting armed conflict in Ukraine (Szulecki and Westphal, 2018). After the annexation, Western countries imposed sanctions on the Russian oil sector (Kuzemko et al., 2016). In May 2014, the "Communication from the European Parliament and the Council" published a European Energy Security Strategy. It first and foremost aimed to increase the EU's capacity to overcome a major gas disruption during the winter of 2014–2015, but also included seven other pillars, such as strengthening emergency and solidarity mechanisms, moderating energy demand, increasing energy production in the EU, and diversifying supplies (EC, 2014). Despite this, the EU's energy import dependence grew and peaked five years later, in 2019 (Figure 3.2).

One explanation for the ineffectiveness of European energy security policies may be the lack of coherence between the economic and security policies of the EU and some of its member states. For example, Chapter 6, outlining the Finnish case, shows how energy has been addressed principally in economic terms prior to 2022, ignoring security. The same has occurred, for instance, in Germany, perhaps even in a stronger manner. Another reason is that the EU energy union context is characterized by divergent national energy security interests and differing energy policy strategies, such as strong advancement of renewable energy in Germany and Denmark, active resistance to energy transition in Poland, and lack of favorable conditions for a transition in Hungary and Romania (Mata Pérez et al., 2019). Achieving policy coherence within the EU energy policy domain itself is not easy, which makes coherence between EU energy and security policies even

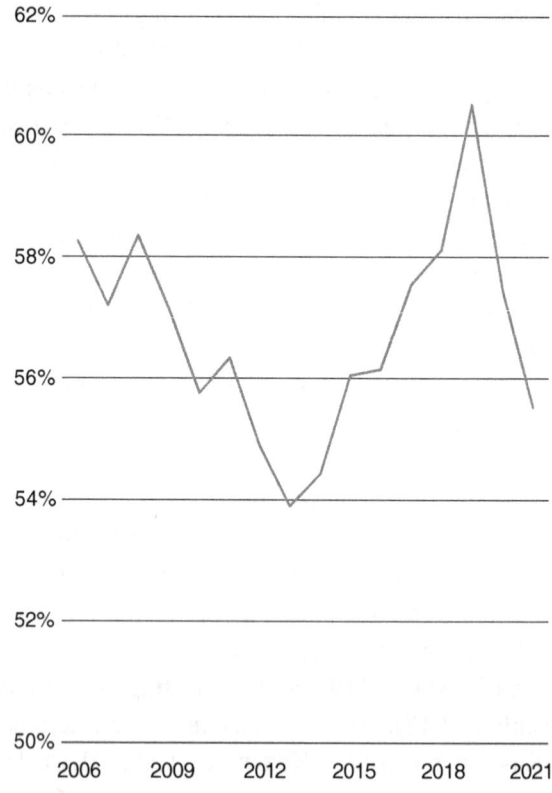

Figure 3.2 EU energy import dependence, 2006–2021.
Source: Eurostat (2023).

more difficult. The EU energy union policy has tried to combine policy goals for security of supply with climate change mitigation and energy market liberalization across the EU with, in hindsight, limited success (Strambo et al., 2015).

Whereas many Eastern European countries have sought to reduce energy dependence on Russia, Germany wanted to construct the Nord Stream pipelines (Bocse, 2020). During 2011–2022, these pipelines supplied natural gas from Russia to Germany via the Baltic seabed, avoiding transit countries. The first construction agreement was signed in 2005, but was strongly opposed by Poland and the Baltic States (Heinrich, 2018). Mata Pérez et al. (2019, p. 1) have effectively described a "multi-speed energy transition," where the Eastern European member states are largely driven by security (of supply) concerns and the Western and Northern states, prior to 2022, by business opportunity and decarbonization.

Considering the events of 2022, it seems that European energy security policy has not been effective enough in preventing natural gas disruptions or developing alternative strategies to produce enough heat and power to overcome any major gas disruptions – such as the one following Russia's attack on Ukraine in

February 2022, the subsequent termination of gas flow via Nord Stream 1 and 2 pipelines, and gas leaks from those pipelines in September 2022 when not in operation. In spring 2022, the European Commission launched the RePowerEU plan to reduce dependence on Russian fossil fuel exports as quickly as possible and fast-forward the energy transition. It outlined both short-term and medium-term measures. The former included, for example, common purchases of gas, liquefied natural gas (LNG), and hydrogen, as well as rapid rollouts of wind and solar projects. The latter comprised, for instance, strengthening industrial decarbonization, increased energy saving ambitions, and a jump in the EU renewables target for 2030 from 40 to 45 percent. The investments made in new LNG infrastructure and related contracts with non-European countries, such as the US, create risks of new path dependencies that are slowing down climate change mitigation, as well as creating new geopolitical and geoeconomic ties. The EU's pursuit of energy security and the aftermath of Russia's attack to Ukraine in 2022 show that energy security cannot simply be regarded as a national issue and that a pan-European approach is needed. This, however, is not an easy task, because European countries have very different approaches to energy policy, as we have already shown through several examples and will further develop in this book via its four country studies.

In Chapter 4, I focus more closely on the analytical approach taken in this book to explore the connection between energy transitions and security. I draw from the literature on policy coherence, alongside the literatures reviewed here and in Chapter 2.

# 4
# A Conceptual–Analytical Approach to Examining Security in Sustainability Transitions and Policy Interplay

This chapter outlines the main analytical contribution of this book, drawing from the literatures described in Chapters 2 and 3, as well as the literature on policy coherence and integration explained here. It combines elements of conceptual–analytical frameworks published in scientific articles as part of the research I undertook for this book. However, it also goes beyond these to create a broader framework to address the security inferences of sustainability transitions and the coherence between energy transition and security policies.

The conceptual–analytical framework adopted in this book looks at security in relation to sociotechnical systems and transitions. The key conceptualization behind how transitions are depicted here is the multilevel perspective (MLP) introduced in Chapter 2 (Geels, 2002, 2005b, 2011). The MLP has been criticized due, for example, to its focus on change in technological artifacts and its lack of agency (Genus and Coles, 2008) and ontological assumptions (Shove and Walker, 2010). However, I see it as valuable in depicting how security can be divided into multiple levels: the landscape as the broader context where security affects, in part, the stability or instability of the sociotechnical energy regime; security as a sociotechnical regime itself that engages in multiregime interaction (including policy coherence) with the energy regime; and the range of positive and negative security implications that ensue from regime destabilization or the expansion of niches and niche innovations. I approach changes in niches and regimes via selected processes that may lead to security effects.

Figure 4.1 shows the overall analytical dimensions used in country case studies (Chapters 5–8; also called here "the country chapters"), which will be further explained and elaborated later in this chapter, highlighting the specific focus areas of the book. The framework merges different viewpoints and perspectives to examine sustainability transitions from a security perspective. It is centered around the x-curve of transitions (Hebinck et al., 2022), where the old regime will gradually destabilize and decline and make space for the new one built with the help

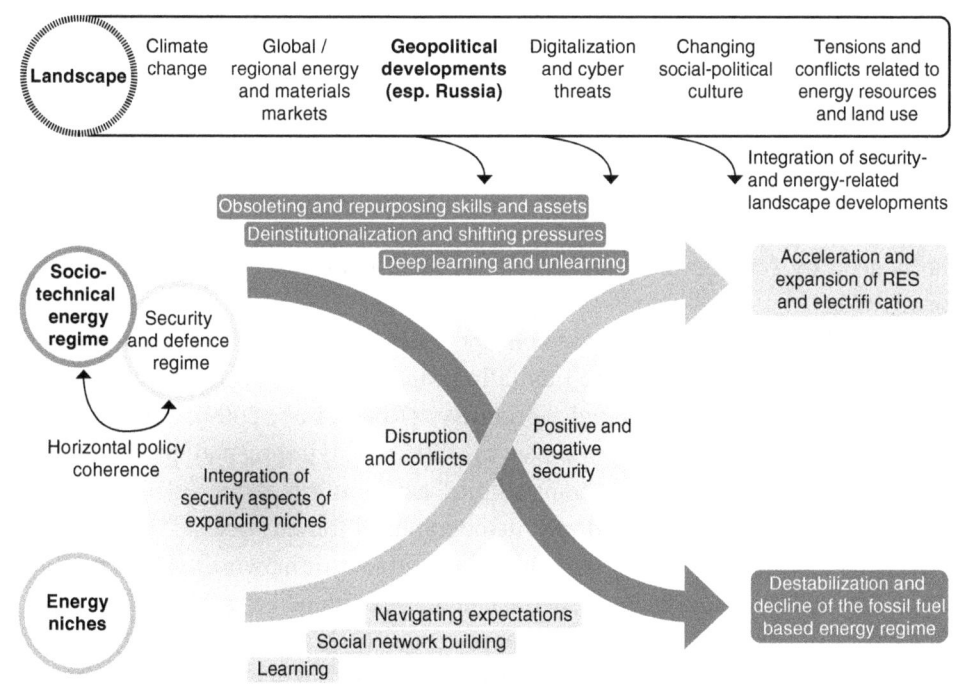

Figure 4.1 Analytical framework for Chapters 4–8.
Sources: Based on adaptations from Geels (2002); Loorbach et al. (2017); Kivimaa and Sivonen (2021); Hebinck et al. (2022); Lazarevic et al. (2022).

of expanding niches. The transition period, where the curves meet, can experience disruption and conflicts.

The framework also draws on the MLP by identifying the three levels of change and how they interlink to security. More specifically, the levels are used to explore: (1) *landscape-level security factors* and how they have been perceived by energy and security experts prior to and post 2022, what potential policy actions may have been taken in the regime level, and whether these actions amount to securitization (see Chapter 2; Heinrich and Szulecki, 2018); (2) *policy coherence between energy (transition) and security and defence policies* at the level of sociotechnical regimes; and (3) the *expected positive and negative security implications of the transition* via the expansion of the renewables niches (and the decline of the fossil fuel-based regime). Nonpolicy-related developments in the regimes are outside the scope of analysis. However, I aim to provide sufficient context in the country chapters in terms of the structure of the energy sectors and the resources available.

In the following, I provide some more detailed explanation on the three focus areas of the conceptual–analytical framework.

## 4.1 Security as Part of the Sociotechnical Landscape for an Energy Regime

The landscape level is the broad context that influences sociotechnical regimes and sustainability transitions. Berkhout et al. (2009) talk about the landscape as the selection environment that contains political, economic, and institutional contexts and conditions for both niches and regimes. It cannot be directly influenced by specific niche actors and regime actors the same way as niches or regimes.

The problem with the landscape concept is that several different types of issues or elements have been described as falling under this conceptualization. These include, for instance, values and worldviews (Rock et al., 2009), scientific paradigms, social movements (Smith et al., 2010), environmental problems, the phenomenon of globalization, transnational actors (Grin et al., 2010), political ideologies, macroeconomic patterns, demographical trends (Geels, 2011), culture (Geels and Verhees, 2011), overarching institutional frameworks (Upham et al., 2014), and natural hazards, wars, and pandemics (Huttunen et al., 2021).

The landscape is not fixed, and it experiences both slowly moving long-term developments and more short-term, or even abrupt, changes. For example, climate change can be depicted as a long-term landscape development, whereas the initial phases of the Fukushima nuclear disaster, the Russian attack on Ukraine, and the COVID-19 pandemic can be depicted as more sudden changes or landscape "shocks."

From the security perspective, the landscape is an extremely relevant transition studies' concept. Pressures threatening geopolitical, environmental, human, or cyber security are quite evident at the landscape level. Regarding the geopolitical dimension, sociotechnical energy regimes have seen landscape changes in the positioning of major states in terms of global alliances or military actions, which have influenced cross-country energy flows and security of supply. The war in Ukraine instigated by Russia in 2022 is an example of how war efforts have led to the energy supply from Russia to Europe being cut off, and changes are envisaged in both energy alliances between countries and physical infrastructure development.

Other security-related landscape factors include, first, the increased risk of cyberattacks, which is heightened as societies become increasingly digitalized. Second, planetary environmental problems that threaten both climate and environmental security and to which energy regimes need to respond. Third, changes related to the increase of extremist right-wing movements and populism, which both connect to human security and influence the degree of landscape-level support for zero-carbon energy transitions. Fourth, globally increasing energy demand

and scarcity of resources as landscape developments also influence security of supply in local, regional, and national energy regimes. When landscape developments or pressures are depicted as threats, based on security studies we should ask what is the perceived "referent object" (see Chapter 2) that is to be secured against such a threat? Is it the energy regime, the unfolding energy transition, the state more broadly, humans, society at large, or the planet?

The landscape can be seen from a global and a more local perspective. I argue that the landscape is effectively formed by both physical developments, such as natural disasters, but also social constructs. Consequently, the ways in which relevant actors *perceive* developments in the wider environment forms the landscape. Antadze and McGowan (2017) argued that landscape developments are interpreted by actors (with agency) for the use of niches and regimes. The country cases of this book show, for instance, that while Russian military and energy developments are a part of a broad landscape context for other countries' energy regimes, they have been – at least prior to 2022 – differently interpreted as landscape pressures by actors in different countries and in different regimes.

The sustainability transitions literature has paid less attention to the landscape than to niches and regimes and, hence, conceptual specifications are limited. Some insights have nevertheless been provided. For example, three different temporal elements for the landscape have been suggested by Van Driel and Schot (2005): (1) factors that do not change or change very slowly, such as the climate; (2) rapid external shocks, such as wars or oil price variations; and (3) long-term changes in specific directions, such as demographical trends. In the 2020s, we have seen a remarkably high number of rapid external shocks influencing energy regimes. These include the COVID-19 pandemic, an increased number and scale of extreme weather events from climate change, and the war conducted by Russia in Ukraine and the resulting implications of this on wider European developments.

The key importance of the landscape concept is its influence on niches and regimes. Frank Geels (2011) makes a distinction between stabilizing and destabilizing landscape influences. Relatively stable landscapes can reinforce existing regimes (Smith et al., 2010). This is visible, for example, in the relatively slow changes in sociotechnical energy regimes in the past. Destabilizing landscape influence can, in turn, be associated with the disruption of sociotechnical systems and their technological, market, policy, or behavioral dimensions (Kivimaa et al., 2021).

The boundaries between landscape and regime are somewhat blurry. For example, while technologies are typically addressed as part of regimes, Rip and Kemp (1998) argued that some technologies are also elements in the landscape, providing an example of motorcars because they have had such a profound influence on broader societal rules and cultures (e.g., perceptions of freedom and cultural necessity). The same could be said for many digital technologies, such as computers and cell phones. In effect, landscape is determined based on what regime is in focus and how it is defined. For example, it may be that renewable energy technologies will have as great an influence on the landscape in the future as motorcars had in the past across the society.

Whereas, generally, regimes are not seen as being able to change landscape factors, Smith et al. (2010) argue that over long periods the creation of new regimes can affect broader landscape developments – describing the examples of the developments of aeromobility and communications technologies affecting globalization. In the context of energy and security connecting to the literature on geopolitics (Chapter 3), zero-carbon energy transition can also shape the broader landscape, such as international relations between states and global stability, affecting different sociotechnical regimes.

The landscape differs according to the perspective of different sociotechnical systems or geographical locations The natural environment, culture, economies, and populations mold the landscape in specific places, regions, or internationally (Rock et al., 2009). Therefore, the landscape for the sociotechnical energy regime and its transitions differs from the viewpoints of different countries or regions. The boundary between the regime and the landscape can be analytically set based on the focus of each study and its scale (e.g., a local, regional, national, or international sociotechnical regime).

Rock et al. (2009) proposed that the sociopolitical part of the landscape is composed of institutions and values that guide the economy. Actors' values are regarded as fairly permanent and, thus, as part of the landscape (Bögel and Upham, 2018), while moves toward more altruistic, biospheric, or postmaterial values would benefit sustainability transitions (Huttunen et al., 2021). Practices conducted by actors link to more general social norms and values of the landscape (Bögel and Upham, 2018; Laakso et al., 2020). Hence, gradual shifts in values and the prioritization of values are important determinants in how the landscape also affects transitions. For example, the energy transition requires environmental values to be prevalent, while longer-term economic values are often employed to convince actors of the benefits of energy transitions. In contrast, short-term economic values can and have often slowed down sustainability transitions. Geopolitical or "realist" values that focus on security of supply, strategic alliances, and military power are also present, but to differing degrees in different countries (Kuzemko et al., 2016).

Landscape pressures can be distinguished as either unintentional or intentional (Morone et al., 2016). Unintentional pressures are, for example, the advancement of climate change or changing demographics. Intentional pressure can be created via

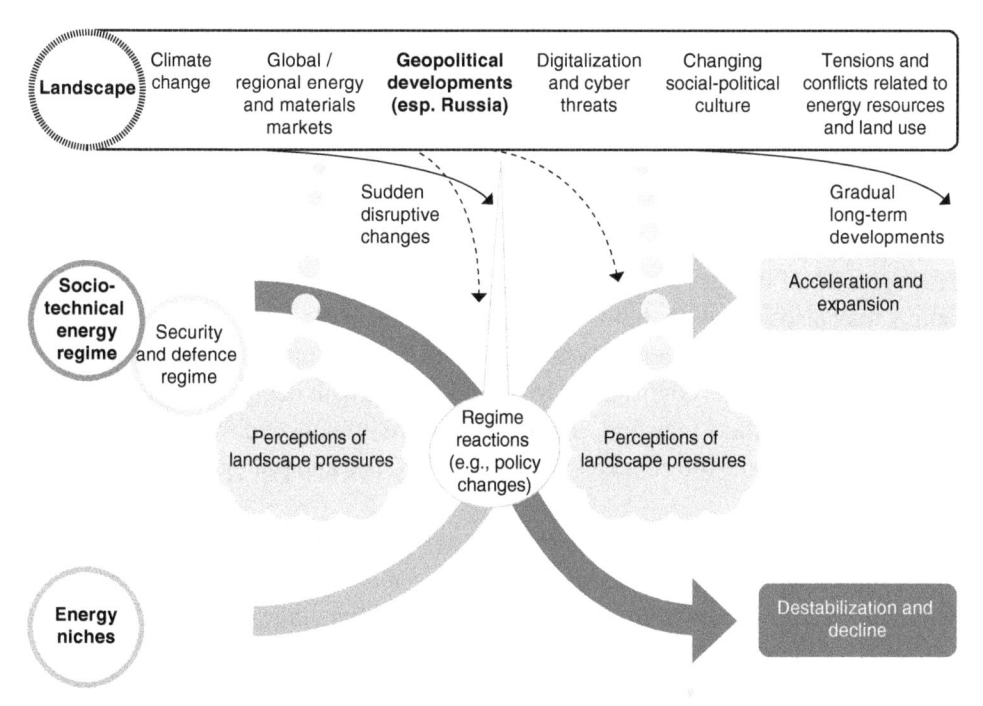

## 4.1 Security as Part of the Sociotechnical Landscape

Figure 4.2 Landscape in focus.
Sources: Based on adaptations from Geels (2002); Loorbach et al. (2017); Kivimaa and Sivonen (2021); Hebinck et al. (2022); Lazarevic et al. (2022).

large-scale institutional changes and political mechanisms, such as oil embargoes or climate change conventions. Both intentional and unintentional pressures and developments are subject to actors' interpretations and perceptions. For instance, the risks posed by geopolitical developments instigated by Russia were seen as more or less significant by different actors before 2022 (Kivimaa and Sivonen, 2023). The same goes for climate change. Despite the widespread scientific consensus on the realization of climate change and the threats it poses, some states, politicians, or economic actors have interpreted, for example, related extreme weather events or Arctic ice retreat as less concerning than others. Therefore, the perceived scale and urgency caused by landscape-level developments on sociotechnical transitions varies.

Figure 4.2 depicts the part of the analytical framework of this book that is focused on the landscape and, in particular, interpretations of that landscape. Developments can be depicted as gradual or more sudden. In the country cases (Chapters 5–8) focus is placed on how these countries and their energy and security experts have perceived developments pertaining to Russia as a landscape pressure for the energy transition, alongside some other key concerns for energy policy. Empirical insights on the perceptions of expert actors before and after 2022 are delivered.

## 4.2 Policy Coherence at the Regime Level: Interplay of Energy Transition Policies with National Security and Defence Policies

Public governance plays an important role in sustainability transitions. Government interventions in the form of public policy can facilitate transitions by setting goals, targets, and specific policy interventions or policy mixes to support changes. The EU Green Deal is a good example of such policy interventions. However, public policy may also frequently hinder transitions, for example by preventing the diffusion of niche innovations and subsidizing or otherwise supporting an unsustainable sociotechnical regime, contributing to its lock-in and path dependence. Policy contradictions or conflicts may undermine the positive influence of transition-oriented policies, which make the concepts of policy coherence and policy integration areas of interest here.

Public governance has traditionally been defined in terms of a unitary state with vertically integrated policymaking and implementation, that is, the policy cycle, while the theory of new public governance posits that state is actually disaggregated and policymaking and implementation at least partly disconnected (Osborne, 2006). National-level public governance in Western countries is typically organized so that a single or multiparty government, formed of elected parliamentary politicians, designate ministers to lead ministries. The composition and number of ministries varies among countries. For instance, climate and/or energy policy can be allocated under their own ministry or be part of a broader ministry, often the Ministry of Economic Affairs. Administrative sectors are formed of ministries and agencies that typically implement the policies set in ministries.

Ministries are often established long-term institutions. Thus, they have long traditions and, frequently, adopt specific worldviews that influence their policymaking practices and objective setting. For instance, ministries of defence have usually adopted a realist and geopolitical worldview. This means focus on state security by military means and, with respect to energy, on aspects such as security of supply and strategic alliances (Kuzemko et al., 2016). Ministries of economic affairs tend to orient toward political liberalism and free market-based worldviews. This implies, for instance, free market trading and limited government intervention in market operations (Kuzemko et al., 2016). Therefore, energy policy subsumed under a ministry in charge of economic affairs is less prepared for geopolitical threats. Ministries in charge of the environment are typically oriented toward environmental perspectives. The worldviews of energy or environmental ministries may also at times connect to socialist perspectives, which, according to Kuzemko et al. (2016, p. 14), relate to "greater equity in the distribution of wealth," affordable electricity prices, and societal well-being as primary objectives over economic profits. Another potentially crosscutting perspective influencing worldviews is a technological one, which may be combined with a technocratic approach toward, for instance, energy or environmental policymaking and implementation. The diversity of worldviews in different administrative sectors make coherent policy-making difficult, while dominant party political views also influence the degree of policy coherence.

Public policy can be described in terms of policy objectives and instruments, as well as processes for setting up and implementing these. The sets of objectives, instruments, and processes influencing a given policy issue, for example building energy efficiency, or a given domain, for example energy policy, can be called policy mixes. Policy mixes have been described as complex arrangements that have gradually formed over the years (Kern and Howlett, 2009) and that exist in a messy multilevel and multiactor reality (Flanagan et al., 2011). This means that "ideal" policy mixes vary from place to place and sector to sector. Rogge and Reichardt (2016) have argued that policy processes are an important part of policy mixes, because policy preparation processes influence how policies are designed and redesigned and implementation processes may, for instance, suffer from political resistance or poor implementation. Indeed, such implementation deficits have been observed in relation to energy efficiency policies (Kivimaa et al., 2017). The process dimension of the policy mix is connected to policy coherence and integration because, for example, policy coordination structures in place between different ministries and agencies influence how broader policy mixes in the interface of energy (transitions) and security are designed and implemented. Broader conceptualizations of policy mixes also include governing organizations and their institutional developments (Kivimaa and Rogge, 2022). Changing institutions and organizations are longer-term processes with potentially crucial impacts on the advancement of energy transitions and on achieving more coherent energy transition and security policy mixes.

Policy contradictions and conflicts may occur both within specific policy mixes (e.g., those designed to govern the national energy production and supply) and between different policy domains across various sets of policy mixes. Alternatively, policies within and across domains can be complementary (i.e., no contradictions) or even seek policy synergies to bring policy objectives and instruments more into alignment. Policy conflict can be defined as a situation where two policies together achieve less than they would separately (Howlett et al., 2015). This occurs when policies give contradictory signals to policy recipients in terms of actions. An example from the energy domain is when renewable energy subsidies aim to advance the diffusion of renewable energy technologies by making them more competitive with fossil fuels, while at the same time fossil fuel subsidies undermine the effect of renewable energy subsidies. This kind of situation can also be described as policy incoherence (Huttunen et al., 2014). Synergies go beyond complementarity, that is, aligned coexistence of policies, in that two policies are synergetic if they together have a greater effect than the sum of both policies' singular effects.

Policy coherence is a concept used to explore policy synergies and complementarities on the one hand and policy conflicts or contradictions on the other. It originates from policy studies (May et al., 2006; Tosun and Lang, 2017) but it is also much applied in practice by organizations such as the OECD and the European Commission. Several types of policy coherence have been described, drawing on different levels and domains of public governance. Carbone (2008) created a typology with four dimensions: horizontal coherence between policy domains, vertical coherence between the EU and its member states, internal coherence as the consistency of objectives and instruments within a policy domain, and multilateral coherence referring to interaction between international organizations. Other conceptualizations of policy coherence also exist and it has been described as an elusive concept, difficult to detect and measure (Righettini and Lizzi, 2022). Nevertheless, in this book, I will try to analyze policy coherence in the case countries.

The analytical framework here focuses on horizontal coherence. Drawing from previous studies, it utilizes the idea of *synergies* and *conflicts* to describe the status of horizontal coherence between objectives, instruments, and the implementation of energy transition policies and of national security and defence policies in the case countries.

The exploration of horizontal coherence is focused on interaction and coordination between two or more administrative bodies or organizations. Such coherence is required to address policy issues, such as climate change, which cut across many administrative sectors (Candel and Biesbroek, 2016). In addition, when a policy issue can create substantial side-effects for another policy domain – for example, energy and security – some coherence is beneficial or even required. Research on policy coherence also emphasizes its processual nature, where policy coherence may weaken or improve over time – it may first get better and then worse again (Candel and Biesbroek, 2016).

Policy integration as a concept is connected to policy coherence. It means the integration of a policy objective, such as climate change mitigation or expansion of renewable energy, to another policy domain, such as security policy. Whereas policy integration does not require coherence or two-way coordination between policy domains, this can benefit the pursuit of policy coherence. When security policy is more attuned to energy questions or the mitigation of climate change it is easier to achieve synergies or complementarities between energy transition policy and security policy. Yet policy integration may also be limited to an isolated functional exercise in a policy domain and not spur interaction between actors from different domains (Kivimaa and Sivonen, 2021).

Policy integration has a history of several decades in the development of European environmental policy. It was developed at the start of the millennium, with different perspectives of integration presented. For example, Lafferty and Hovden (2003) proposed a definition of environmental policy integration including a principled priority of environmental issues over other policy objectives. Nilsson and Persson (2003), in turn, took a learning-based approach in defining and analyzing environmental policy integration, arguing that such integration occurs when (policy) actors meet together and discuss issues. Such learning could, in these instances, occur across political frames or worldviews and their interpretations. Russel and Jordan (2009) distinguished between the approaches of policy integration as normative, organizational, procedural, output-based assessments and reframing. More recent literature has begun to question the feasibility of policy integration in each possible context. For instance, Candel (2021) has argued that policy integration can be costly and deliberation is required regarding when integration is a good use of public resources. He, however, also remarked that not considering policy integration can be dangerous too and can result in disruptive effects in cases of crises. Therefore, policy integration as an idea should not be disregarded without proper consideration.

Certain elements or mechanisms benefit from advancing policy coherence and integration and can also function as analytical evidence of the presence or absence of coherence and integration (see Kivimaa, 2022a, for details). While many of these mechanisms, such as shared visions or specific plans to improve coherence, operate at the level of administration, the political level is important too – albeit rather sparsely addressed. Tosun and Lang (2017, p. 559) argue that political leadership and parliamentary committees are important for policy integration, noting that "political dynamics have not been systematically explored within the literature on horizontal governance." Also, Runhaar et al. (2018) emphasize political commitment as an important explanatory factor. Jordan and Lenschow (2010) note that lack of political will is associated with mere symbolic actions on policy coherence and integration, particularly by right-wing governments. In effect, improving policy coherence or integration may require a shift in some dominant political frames (Candel and Biesbroek, 2016). The 2022 security and energy crisis in Europe may have created such a shift in the dominant frames for many countries and the EU more broadly.

Specific elements or mechanisms for policy coherence, proposed in the literature, include visions (May et al., 2006) and comprehensive frameworks (e.g., common strategic objectives and instrument mixes) shared across policy domains (Furness and Gänzle, 2017). These can be implemented as new policy strategies. Within and across public organizations, coherence and integration can be advanced by setting up new executive agencies (Tosun and Lang, 2017), creating means of coordinating between sectoral administrations, promoting specific plans for coherence, allocating staff and financing (Runhaar et al., 2018), and evaluating and reporting on coherence and integration. Independent working groups or science panels may also be used (Mickwitz et al., 2009). While they may not guarantee the presence of synergies or the absence of conflicts these elements can be interpreted as signs of attempted coherence. From a transition perspective, they may nevertheless be useful, as transitions have been argued to benefit from constructive and open tensions among actors. Thus, more explicitly recognizing policy conflicts is a start to exploring connections among diverging worldviews, interests, and perceptions. However, policy incoherence or lack of sufficient policy coherence or integration are common. Conflicting interests or lack of access to knowledge and advice results in poor integration and incoherence, visible, for example, as conflicting policy statements and objectives (Runhaar et al., 2020). Cultural and cognitive frames of policymaking influence how and whether policy coherence and integration happens (Jordan and Lenschow, 2010).

In the context of policy coherence for energy and security, an important additional feature is the dynamics of securitization (see Chapter 2). In essence, securitizing energy policy could mean, at least in some cases, a principled priority of security over other policy objectives. This could be translated as extraordinary energy-policy measures for security reasons that are not part of established political or policy practices, allocating more power from the ministries to the agency level in decision-making related to energy and security and/or hiding information from the public eye (Heinrich and Szulecki, 2018). Securitization of energy policy could lead to improved coherence between energy and security policymaking, but the viewpoint taken would determine the effects on zero-carbon transition in terms of contradiction or synergy. For energy transition policies to be aligned with security policies would require acknowledging environmental and climate security as central parts of security policy and the pursuit of securitization.

Figure 4.3 shows the part of the analytical framework oriented toward exploring horizontal policy coherence between energy policies and security and defence policies from a transitions perspective. The policy domains are in interplay, not in a static world, but in a world where the energy transition is advancing and where landscape-level developments are taking place, creating new pressures for national security and the energy sector. The analysis focuses on synergies and conflicts/contradictions, administrative coordination between the domains, and the existence of potential coordinating elements. In addition, it aims to identify how security aspects of expanding energy niches as well as new landscape developments have been integrated into the nexus of energy and security policymaking.

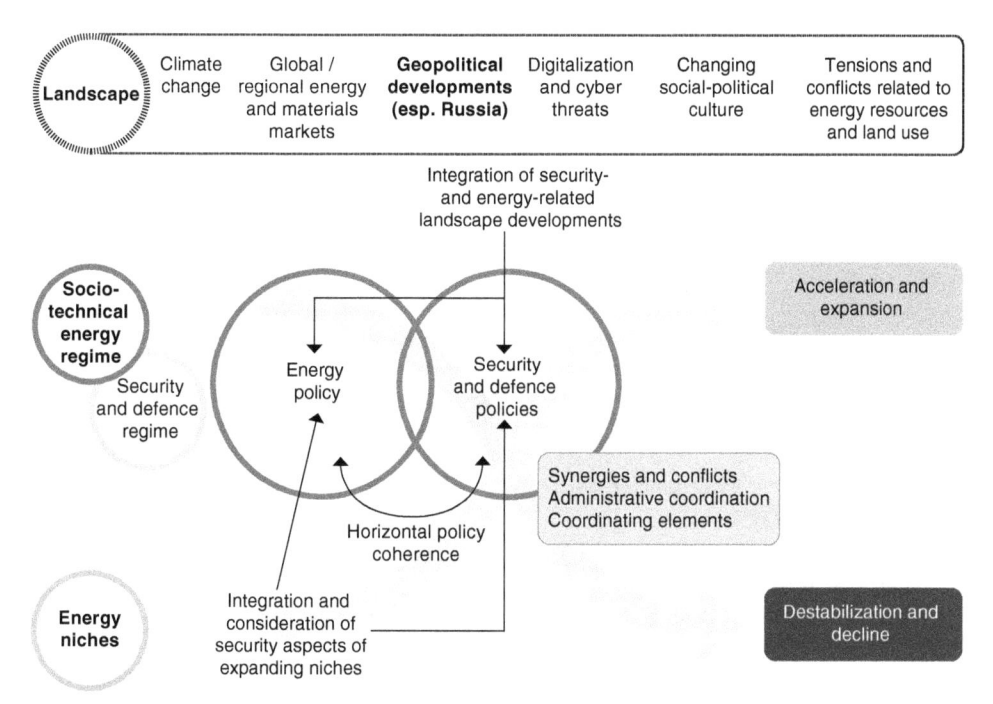

Figure 4.3 Coherence between energy transition and security and defence policies at the sociotechnical regime level.
Sources: Based on adaptations from Geels (2002); Loorbach et al. (2017); Kivimaa and Sivonen (2021); Hebinck et al. (2022); Lazarevic et al. (2022).

## 4.3 Security in Change Processes: Niche Expansion and Regime Decline

The final part of the analytical framework is focused on the positive and negative security implications of the unfolding energy transitions (Figure 4.4). It draws on the concepts of positive and negative security, introduced in Chapter 2, and on processes of niche-building from sustainability transitions literature. It also proposes new processes for regime decline (Kivimaa and Sivonen, 2023), drawing from literature on regime destabilization and decline. The aim is to note how the different case countries have explored the security implications of these processes in public policy development (strategies, policy actions) and why the countries may have differing perspectives on this. In country chapters, this topic is mainly addressed via specific cases where security connects to a certain energy technology or development.

I now briefly describe the key analytical components used, drawing from a scientific paper in which they were first used (Kivimaa and Sivonen, 2023). Table 2.1 in Chapter 2 described the established processes of navigating expectations, social

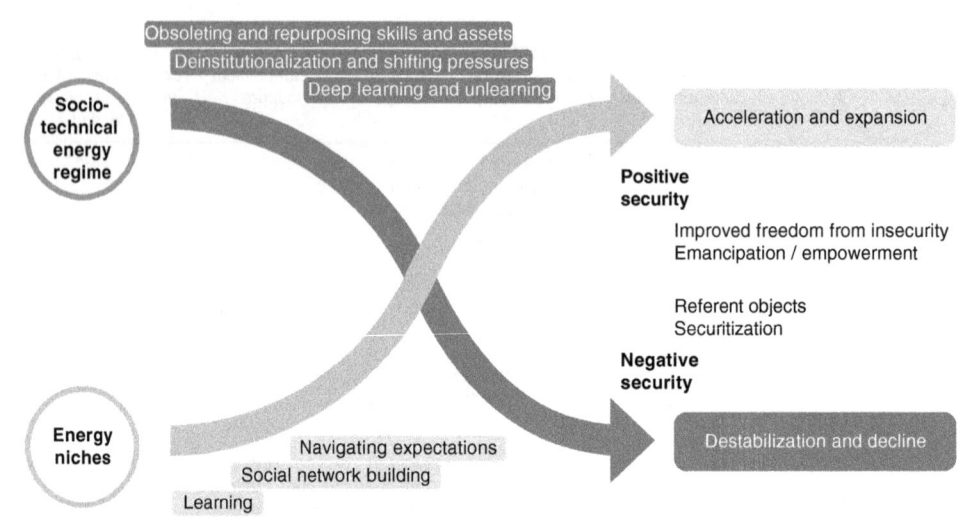

Figure 4.4 Framework for analyzing the positive and negative security of transition processes.
Source: Elaborated from Kivimaa and Sivonen (2023).

network-building, and learning, which have been used to delineate the development of new niches but can also be applied in the context of assessing their security implications. Because similar established processes for regime decline do not exist, new processes that could be used to describe regime decline and explore the security implications of fossil fuel phaseout are proposed (Table 4.1).

It has long been argued that disruptive innovation leading to regime destabilization may initiate processes that reduce the value of existing skills, knowledge, competences, and resources (Abernathy and Clark, 1985; Kivimaa and Kern, 2016) and weaken the flow of resources to previous core technologies (Turnheim and Geels, 2013). New research, however, shows that incumbent actors in the energy sector are increasingly seeking to repurpose their resources to new technological and market contexts (Mäkitie, 2020). This process is described as *disruption to and repurposing* of skills and assets (Kivimaa and Sivonen, 2023, p. 1).

In the energy security context, I regard the second process of regime decline to be *unlearning and deep learning*. Unlearning has been used to refer to processes that question and reject taken-for-granted values, norms, and beliefs (Feola et al., 2021) linked to incumbent power structures (Stirling, 2019), as well as discarding ineffective habits and practices alongside established mental models (Van Mierlo and Beers, 2020; Van Oers et al., 2023). Deep learning, in turn, indicates experiential social learning about the pressures for change for established regimes and creating new in-depth knowledge about change to come (Ghosh et al., 2021).

Table 4.1 *Processes of disruption to and repurposing skills and assets, unlearning and deep learning, and deinstitutionalization and shifting pressures in regime decline*

| Regime decline process | Grounding in literature |
| --- | --- |
| Disruption to and repurposing skills and assets | Disruptive innovation and resulting regime destabilization processes create changes, whereby existing skills, competences, knowledge, and resources may become reduced in value and, in extreme cases, obsolete (Kivimaa and Kern, 2016). In an industrial context, this means that the value of incumbents' expertise and other factors of production reduces significantly (Abernathy and Clark, 1985). Destabilization is argued to weaken the flow of resources into the reproduction of regime elements such as core technologies (Turnheim and Geels, 2013) and financial resources (Rosenbloom and Rinscheid, 2020). However, recent research, for example, in the COVID-19 pandemic context, also shows that actors can quite rapidly repurpose their skills to new types of commercial operations (Nemes et al., 2021). For instance, Norwegian oil and gas industry companies have sought new corporate ventures in offshore wind, mainly because they are able to repurpose their existing resources (e.g., technological and market expertise) to this new form of power (Mäkitie, 2020). |
| Unlearning and deep learning | Unlearning and deep learning are connected to processes that destabilize existing sociotechnical regimes. Unlearning is a process that results in discarding old obsolete practices and ineffective habits (Van Mierlo and Beers, 2020), established routines, and mental models (Van Oers et al., 2023) via consciously not thinking or acting in old ways (Stenvall et al., 2018). It has been described as a continual and reflexive process of identifying how our conceptualizations of the world are unself-consciously bounded (Lawhon et al., 2016). It is about rejecting and questioning taken-for-granted values, norms, and beliefs (Feola et al., 2021) that are often associated with incumbency and the structuring of power (Stirling, 2019). Unlearning means accepting a certain risk and uncertainty about the future regime and, hence, connects to deep learning, that is, "experiential social learning" about challenges facing the extant regime and constructing new in-depth knowledge about the changing system dynamics (Ghosh et al., 2021). |
| Deinstitutionalization and shifting pressures | Deinstitutionalization is a process where legitimacy is eroded in the context of shifting social, political, and functional pressures, implying changes in underlying interests and power relations and in structures of leadership and authority, and reducing cultural consensus (Novalia et al., 2022). Key actors in (de) institutionalized structures may be replaced organically or in response to deliberate attempts (Kivimaa and Kern, 2016; Turnheim and Geels, 2012). Dominant actors may lose influence and legitimacy when markets decline, and when value chains and networks break up with weakening expectations connected to changing landscape pressures (Markard et al., 2020), while actors also resist this process by seeking renewed roles in the new system that relegitimize their position (Mäkitie, 2020). |

*Source:* Kivimaa and Sivonen (2023).

Further, regime decline can be connected to *deinstitutionalization and shifting pressures*. In deinstitutionalization, legitimacy is decreased when social and political pressures change, resulting in changes in fundamental structures of leadership and authority, power relations, and interests (Novalia et al., 2022). Niche development and regime decline processes dynamically influence each other and may even overlap. Hence, they cannot be considered as mutually exclusive, but rather as complementary, processes.

In the country chapters, I focus on selected cases related to niche development and regime decline that appear important from the perspective of security. Identification of such cases was not always easy, because many previous niches (such as wind and solar power) have increasingly become a part of regimes. In turn, some energy sector developments, especially those related to nuclear energy or hydropower, cannot be described in terms of either niche development or regime decline and are, rather, nondeclining parts of incumbent energy regimes – but their contexts change when transitions are in place in the broader energy system. The cases explored in the country chapters include, for example, wind power and defence air surveillance radars (Estonia and Finland), security of hydropower infrastructure (Norway), oil shale phaseout (Estonia), peat phaseout (Finland), and nuclear security (Scotland).

This ends the first, conceptual, part of this book. Part II presents empirical analyses of the four country case studies and shows how security and defence intertwine with energy policy questions and transitions in these countries.

# Part II

Empirical Case Studies

# Part II
Tropical Case Studies

# 5

# Estonia

*Long-Term Energy Independence and Oil Shale*

Estonia is a small European country with a population of 1.4 million and an area of circa 45,000 square kilometers. It is one of the three Baltic countries located beside the Baltic Sea. It has a land border connecting with Russia and Latvia and it is only 80 kilometers by sea from the capital of Finland, Helsinki. Only some thirty years have passed since Estonia regained independence in 1991, when the Soviet Union dissolved. It had already declared independence much earlier, in 1918, after World War I, but was occupied by the Soviet Union from 1940 until 1991. In 2004, Estonia became a member of the EU and the North Atlantic Treaty Organization (NATO).

Estonia's energy policy has been heavily oriented toward national security concerns and attempts to break connections with the Russian energy system. The general orientation has been to reduce the geopolitical threat of Russia, that is, negative security. The development of oil shale has been important to Estonia in its disconnection pursuits from the Russian energy supply and, hence, this has slowed down Estonia's zero-carbon energy transition. Yet the location of the oil shale production has been close to the Russian border, reducing the security of this energy type due to potential Russian intervention. Gradually, Estonia has stopped commercial electricity trade with Russia, despite still-existing infrastructural connections, and has aimed to cover consumption by using domestic sources while also expanding energy interconnections with the EU (Kama, 2016). This decision made Estonia the EU member state with the lowest energy import dependence. Estonia's reliance on energy imports reduced from 34 percent in 2000 to less than 5 percent in 2019.[1] From 2010, Estonia became a net exporter of electricity generated from oil shale. However, by 2020, the country's energy import dependence had increased again to 11 percent (Statistics Estonia, 2023), partly due to the emission reduction requirements of EU climate and environmental policy.

---

[1] https://ec.europa.eu/eurostat/cache/infographs/energy/bloc-2c.html (accessed February 4, 2022).

Since the early 1990s, this country has experienced a relatively rapid development, from being part of the "laggard" Eastern European Bloc to a nation with advanced digital technologies. Despite the general innovativeness of Estonia, its energy system has been largely tied to an old fossil fuel-based regime, in particular oil shale, and dependence on Russian electricity and gas networks. Oil shale is "an energy-rich sedimentary rock that can either be burned for heat and power generation or used for producing liquid fuels" (IEA, 2019, p. 11); it has a carbon intensity similar to coal but is less efficient as a fuel due to a lower share of organic matter. The mining of oil shale began officially in 1918 when a government decree made open pits the responsibility of a particular department in the Ministry of Trade and Industry (Holmberg, 2008). Estonia's energy supply has relied almost entirely on domestic oil shale since 1991.

Oil shale consumption dropped dramatically, over 40 percent, during the 1990–1995 period, in connection with Estonia gaining independence.[2] However, after 2000 there was a gradual increase in total oil shale consumption, peaking at 15 million tons in 2013, a 42 percent increase on 2000 (Statistics Estonia, 2023). This was followed by a gradual decline, and then sudden drops to 5.9 million tons in 2019 and 3.1 million tons in 2020 (see Figure 5.1). This decline in the share of oil shale was caused by rising $CO_2$ emission quota prices, making this type of energy less competitive in the European electricity markets (Vasser, 2021). In addition, some large oil shale plants had reached their full operational age. However, the data also shows a rather rapid increase of oil shale consumption more recently, in 2022, when the energy crisis hit Europe following the Russian invasion of Ukraine.

The overall share of natural gas has been low within Estonia's total energy consumption, while the gas used in Estonia was mostly supplied from Russia until 2015. A Russian company, Gazprom, was the main supplier to all the Baltic countries and the only gas supplier to Estonia, meaning these countries were all vulnerable to sudden disruptions (Pakalkaitė and Posaner, 2019). This is illustrated by the gas dispute in 2009 between Russia and Ukraine that left many EU countries that relied on Gazprom with severe shortages (Štreimikiene et al., 2016). Estonia's concerns about Russia's tactics to maintain a monopoly of gas transit to Europe has resulted in it denying the Nord Stream pipeline access to its territorial waters (Crandall, 2014). In December 2014, a new terminal for liquefied natural gas (LNG) was opened in Lithuania, which allowed Estonia to diversify its supply from Gazprom (Pakalkaitė and Posaner, 2019). A new gas pipeline, Balticconnector, was opened between Finland and Estonia in 2020, connecting gas markets in the two countries, with the aim to create a regional Baltic–Finnish gas market. The

---

[2] Data source: www.iea.org/countries/estonia (accessed February 2, 2021).

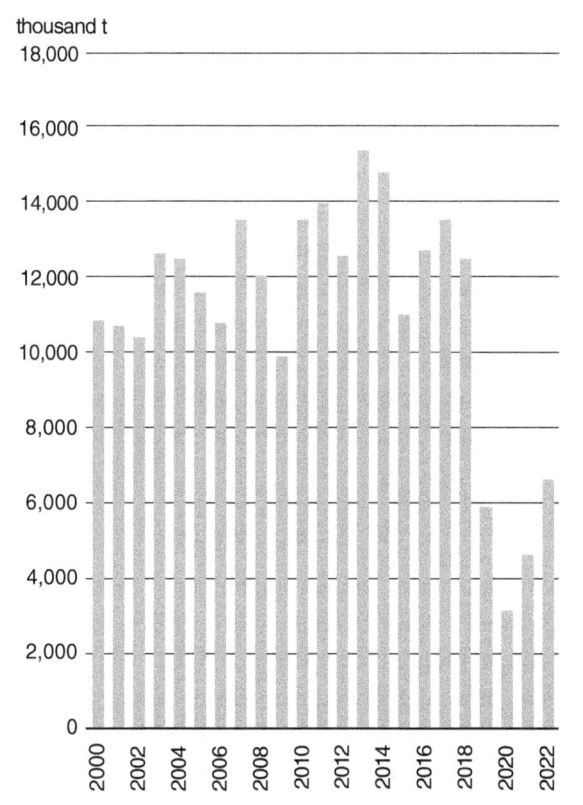

Figure 5.1 Total consumption of oil shale (thousand tons), 2000–2022.
Source: Statistics Estonia (2021).

pipeline was massively damaged in fall 2023 halting supply. Investigation to this incident was inconclusive regarding intentionality of this damage, with the likely cause being damage by a ship anchor. Almost 40 percent of natural gas is used in district heating, but, due to increasing natural gas prices, district heat producers are increasingly considering local renewable fuels (MEAC, 2022).

Estonia met the EU target for renewable energy, 25 percent, in 2011, primarily through biomass and wind power, although the former caused degradation of biodiversity. The shares of renewable energy and combined heat and power (CHP) in the country's electricity production have rapidly expanded. In 2021, renewable energy amounted to 38 percent of total energy and 29 percent of electricity production, while oil shale still made up over a half. Of renewable electricity production in 2022, about half was from wood chips and waste, a quarter from wind power, and 20 percent from solar power (Statistics Estonia, 2023). The expansion of the wind power niche has been limited owing to concerns related to the effects of wind turbines on the defence sector's air surveillance radars. This issue is explored later in this chapter.

This chapter analyzes the first country case study of this book. It creates the context of energy and security regimes in which the interplay of energy and security policies occurs. It then follows the analytical foci presented in Chapter 3: the perceptions of Russia as exerting landscape pressure on energy transitions, policy coherence and integration between energy and security, and positive and negative security aspects related to niche development and regime destabilization via three cases. These are: the phaseout of oil shale, the effects of wind turbines on defence air surveillance radars, and desynchronization from the Russian electricity system. The empirical data analyzed include government energy and security strategies published during 2006–2021 and two rounds of interviews with energy and security experts, between October 2020 and May 2021 and between January and March 2023. These primary sources have been complemented with the literature.

## 5.1 Energy Regime

As outlined, the key sources of the Estonian energy system have been oil shale, imported natural gas, and a range of renewable energies, especially wood chips and waste streams. Unlike in Finland (Chapter 6) and the UK (Chapter 8), Estonia does not have any nuclear power. A government-level working group was established in 2021 to investigate the possibilities of establishing small modular reactors. One complication is that Estonia has no experience of regulating nuclear power so proceeding with this would require drafting new legislation and increasing nuclear expertise.

A developing renewable energy niche is that of offshore wind power, for which there are significant plans. The planned developments include two sites in the Gulf of Riga and near the western island of Saaremaa that would cover 1,700 square kilometers and provide circa 7 gigawatts (Vanttinen, 2022). By 2030, the Estonian government aims that 65 percent of total consumption will be provided by wind power (CPTRA, 2023). Maritime planning was advanced in early 2022 to investigate the effects of offshore wind on defence, shipping routes, and the environment.

Besides renewable energy development and in direct opposition to the aim to decarbonize the energy system, new applications to exploit oil shale and gas are still being made. For instance, the Estonian Ministry for Economic Affairs and Communications has been interested in the shale gas developments in North America, which are a result of substantial shale gas deposits found in the US (MEAC, 2022). In addition, connected to the need to diversify gas supplies, biomethane from local bio-based raw materials has garnered increasing interest. In 2021, Estonia had two biomethane production plants, producing up to 40 percent of the gas used in transport.

## 5.1 Energy Regime

As Estonia is a small country, its administrative structure concerning energy and climate governance is relatively simple. The Ministry of the Economic Affairs and Communications (as of July 2023 the newly established Ministry of Climate) oversees Estonia's energy policy. The Ministry of Finance has some responsibilities regarding budget preparation for the national "Energy and Climate Plan." The Ministry of the Environment (from July 2023 onward the Ministry of Climate) coordinates environmental policy and the management of natural resources. A state-owned company for oil security of supply was established in 2005, which was reorganized as a broader security-of-supply organization akin to the Finnish National Emergency Supply Agency (NESA) (see Chapter 6) in response to the COVID-19 pandemic. A governmental Climate and Energy Committee was established to operate during 2019–2021, with the prime minister chairing this committee, which also had the Minister for the Environment, the Minister for Defence, and the Minister for Foreign Affairs among its members. In 2021, the government established a Green Policy Steering Committee, chaired by the prime minister, to coordinate the implementation of "green policy" across sectors. The ministerial committee is supported by an expert group and makes proposals to the government. In 2023, an expert-based Climate Council was established.

A rather unique feature of Estonian energy policymaking is that there is no agency level. Energy agencies typically regulate and monitor operation of electricity and gas markets as well as emission reductions, or advance energy efficiency. In Estonia, instead, the electricity and gas transmission network operator Elering effectively plays the role of an agency in security of supply. According to an expert interviewee, it is "a quasi-security police or authority." This role is at least partly due to the large role of electricity in Estonia's energy transition, namely desynchronization and offshore wind efforts. In addition, before the 2020s, the security-of-supply organization only addressed oil and not electricity. Some regard Elering's position as a conflict of interest, resulting in little expertise in the public sector:

Sometimes you get the impression that Elering plays the energy agency's role, rather than some big kind of a state agency or energy and climate agency or energy agency or you know whatever you call it, so it's certainly a structural weakness. (Researcher, 2021)

Eesti Energia, a fully state-owned company, is the main producer of electricity. Ninety percent of oil shale-based electricity was produced by Eesti Energia in 2020 (Sillak and Kanger, 2020), but it aims to stop production of electricity from oil shale by 2030 and increase renewable energy production. The company is among the largest employers in Estonia and it is free to make its own investment decisions irrespective of its state ownership (Tõnurist, 2015). The dominance of this state-owned monopoly meant that, until 2014, there was little distinction between the management of the oil shale industry and government energy policy; an attempt to reduce the influence of the industry on energy policymaking was made by transferring the company's ownership from the Ministry of Economic Affairs to the Ministry of Finance in 2014 (Kama, 2016). Nonetheless, it still appears that the company has more influence on Estonian energy policy than policymaking has had on the company.

Several political figures and civil servants in Estonia have held both ministry positions and positions in the energy industry. For example, Taavi Veskimägi has been the Minister of Finance (2003–2005) and the director of the transmission system operator Elering (2009–2023). In addition, at least four individuals interviewed have worked for the energy sector (Elering or Eesti Energia) and the Ministry for Economic Affairs, the Ministry of Foreign Affairs, or the Ministry of the Interior. This is partly explained by the small population of the country, that is, there is a limited number of experts, but also by the nature of informal energy relations and the tightknit relations of the "energy elite."

Another important feature is the relatively late growth of the formalized environmental movement in Estonia. The Estonian Green Party was only established in 2007, despite the fact environmentalism has played an important role in Estonia's independence pursuits since the 1980s (Auer, 1998). The Renewable Energy Council was formed in 2011 by renewable energy entrepreneurs and associations, becoming a vocal advocacy group and proposing a plan for transitioning the country's energy system into a 100 percent renewable energy system (Sillak and Kanger, 2020).

Since 2004, when Estonia became an EU member state, its energy and climate policy has been shaped by the EU. Estonia had already applied for membership in 1995 but, according to an interviewed expert, was unable to join then because its oil shale industry conflicted with EU environmental objectives. Eventually, the EU agreed to give oil shale a temporary status, prolonging the deadline for renovating old power plants to meet EU air-quality standards until 2016, allowing Estonia until 2013 to fully open up its electricity market, and allocating research and development (R&D) support for reducing the environmental impacts of oil shale (Sillak and Kanger, 2020).

Despite these early concessions, Estonia has taken some time to liberalize and decarbonize its energy regime. When the electricity market was fully opened up in 2013, Eesti Energia maintained a significant role in the market. Throughout the years, Estonia has managed to keep large part of the energy production under state ownership, even though this is at least partly contradictory to EU energy market rules. Kama (2016) has noted that "[t]his state-regulated and extremely carbon-intensive sector has shown remarkable resistance to liberalisation, unbundling and privatisation; the three neoliberal modes of energy governance that underpin the policies of the EU, the OECD, and international donor agencies" (p. 836). She continues:

> The government's recurrent flirtations with privatisation further demonstrate how energy security arguments blur the distinction between market and non-market practices in state policy, for conservative and liberal parties alike. Despite two serious attempts to induce private investment capital for replacing the Soviet-era thermal units commissioned over forty years ago, the energy monopoly remains fully owned by the state. (Kama, 2016, p. 842)

Likewise, oil shale has not been addressed to the degree it should have according to EU requirements. When oil prices declined and the price of EU emissions-trading permits rose, the oil shale market became less lucrative. In response to this, the Estonian government lowered environmental taxes to support the industry, justifying this action by claiming the fuel was important for national interests, such as security and employment (Holmgren et al., 2019). In 2019, the International Energy Agency (IEA) noted in its country review of Estonia that "the negative externalities of fossil fuels are currently not sufficiently reflected in the existing tax rates and there is a significant number of tax exemptions and reduced tax rates which are counterproductive for meeting the climate targets" (IEA, 2019, p. 28). An interviewee noted that:

> We still thought that shale oil is going strong and oil shale burning is fine, so, [you saw] even some misguided investments like Eesti Energia putting up a new unit, spending over half a billion euros to continue burning oil shale and so on because the focus was on reducing the $SO_2$ emissions rather than $CO_2$ emissions. (Researcher, 2021)

During 2017–2018, Estonia prepared some key documents for its medium-term energy and climate policy, including the "National Development Plan for the Energy Sector until 2030" and the "Estonian National Energy and Climate Plan 2030," submitted to the EU. Estonia aimed to reduce greenhouse gases 70 percent by 2023 and 80 percent by 2025 by increasing renewable energy (above 25 percent of total energy use), while simultaneously maintaining a high level of energy security by limiting fuel imports, diversifying energy sources and supplies, and connecting with the European electricity market. In the baseline year for emission reductions, 1990, the greenhouse gas emissions were so high, 40,233.79 tons of CO2 equivalent, that it was relatively easy to reduce these by 20 percent by 2020 as required by the EU. The greenhouse gas emissions had dropped almost by half already by 1993 and were on the same level until 2018. Further, Estonia aimed to produce 50 percent of domestic final electricity consumption via renewables by 2030. Wind energy and solar energy for electricity production were planned to be further developed, and biomass use increased for heating. Biofuels and hydrogen fuels were also to be developed, the latter being on the agenda of the Green Policy Steering Committee. The plans were updated in fall 2023. The revised plans show

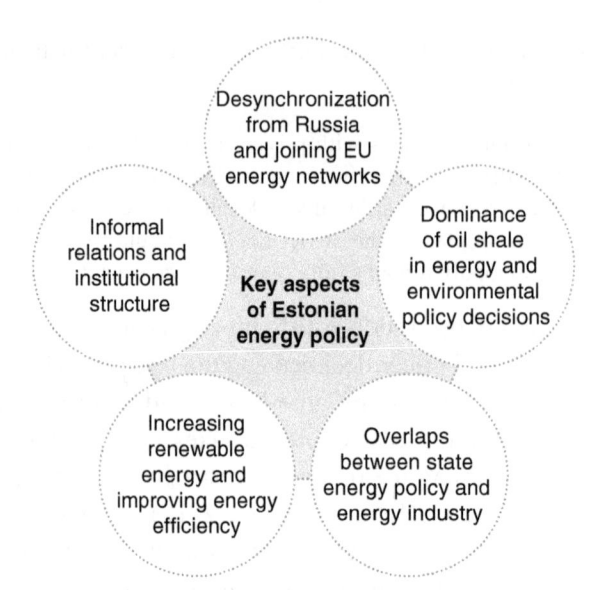

Figure 5.2  Key aspects of Estonian energy policy.

[Diagram showing five circles around "Key aspects of Estonian energy policy": Desynchronization from Russia and joining EU energy networks; Dominance of oil shale in energy and environmental policy decisions; Overlaps between state energy policy and energy industry; Increasing renewable energy and improving energy efficiency; Informal relations and institutional structure.]

an aim to achieve 24 percent reduction in emissions by 2023 compared to 2005, and net-zero emissions by 2050.

Energy independence from Russia has been the cornerstone of Estonian energy policy since its independence. Its EU membership has, however, forced Estonia to change its traditional energy policy and pay attention to the decarbonization of its energy system, which has impacted its energy independence to some extent. This decarbonization process has been slow on both security and economic grounds. Despite the ambitious plans for renewable energy and energy efficiency, oil shale has continued to play an important role in energy policy. The "National Development Plan for the Use of Oil Shale 2016–2030" stated that oil shale will remain an important fuel. The "National Development Plan for the Energy Sector" also mentioned the possibility of new investments in the oil shale sector to increase energy exports. If this continues, it will be hard to lower Estonia's ecological footprint. Figure 5.2 summarizes the key aspects of the Estonian energy regime.

## 5.2 Security Regime

In Estonia, security policy is the responsibility of the Ministry of Defence. The Ministry of the Interior and its agencies play a supportive role, with responsibility for policing, border control, rescue, migration, and immigration management. For information security, different ministries have their own responsibilities, coordinated by the Ministry for Economic Affairs and Communications. The ministries of defence, of the interior, and of foreign affairs also have important tasks with regard to cybersecurity. At the agency level, the Foreign Intelligence Service and Estonian Security Policy Board are important.

Estonia joined NATO in 2004, and this move has had a large impact on its security policy and defence spending. The latter has gradually increased from 1.39 percent of GDP in 2006 to 2.35 percent in 2022. After Russia's actions in 2014 in Crimea, Estonia has become "a meaningful player in NATO's management of renewed Russian security threats" (Studemeyer, 2019, p. 789). The capital city Tallinn hosts the NATO Cooperative Cyber Defence Centre of Excellence. Estonia has developed leading expertise in cybersecurity for NATO, alongside with prioritizing its membership fees to it, even during times of economic austerity (Studemeyer, 2019).

Estonian Defence Forces were reestablished immediately after independence (Piirimäe, 2020). They consist of 4,000 people in permanent readiness and another 4,000 in supplementary reserve. Thirty thousand reservists have also been trained by the Defence Forces (Estonian Defence Forces, 2023).

Security and defence governance are strongly based on the country's history with Russia. Alongside the other Baltic countries, after its independence Estonia sought security by integrating with Western structures: the EU and NATO. The latter led to a revision of security and defence strategies that are reactive to changes in the surrounding geopolitical environment (Piotrowski, 2018). Despite becoming an EU member state, Estonia's foreign policymakers have maintained a traditional realist approach to the East, their security policy "driven by an existential concern about national security," with the hard security guarantees of NATO being a cornerstone of the country's security policy (Raik and Rikmann, 2021, p. 606). These reflect thinking along the lines of "negative security" (cf. Hoogensen Gjørv, 2012). Eurosceptic views have been present all the time but were marginalized until 2015, when the Conservative People's Party of Estonia (EKRE) won 8 percent of the parliamentary seats and later expanded its populist presence in government politics (Raik and Rikmann, 2021).

The security risk from Russian interference is perceived as tangible even after three decades of independence and before Russia's attack on Ukraine in 2022. The elevated perception of risk is not only due to historical factors. It has been influenced by the very slow withdrawal of Russian troops after independence, with some thousand soldiers remaining until August 1994 and an "increasingly aggressive Kremlin policy towards its neighbouring countries" (Piotrowski, 2018, p. 47).

Security policy in Estonia is based on a broad security concept. This means the capability of the state to defend its values and objectives from military and nonmilitary risks. Security policy aims to guarantee independence, sovereignty, survival, and the constitutional order. The key policy documents for security and defence policy are the "National Security Concept of Estonia" (2017) and the "National Defence Development Plan 2017–2026."

Figure 5.3    Key aspects of Estonian security and defence policy.

There is a specific focus on cybersecurity because Estonia is one of the most digitalized societies in the world. Besides Estonians being among the highest internet and mobile broadband users, the country had already installed 100 percent smart electricity metering by 2016, improving network management (IEA, 2019). Cyberattacks from Russia are some of the most concrete threats that Estonia has more recently experienced. Russia considers Estonia as "ideologically unfriendly" and has tried to cause economic disruption by putting pressure on and maintaining tensions between the Russian nationals in Estonia and the Estonian population via cyberattacks (Studemeyer, 2019). Following a sequence of nationwide cyberattacks in 2007, Estonia decided to develop its cyber expertise, including the first governmental cybersecurity strategy in 2008, by providing university-level education in this field, and establishing a "cyber defence league" to act under its own military command in case of war (Crandall, 2014).

The trend is, and this is not surprising, that the cyberattacks are increasing every day. And they are just not joking, but we see they're professional. We see the professionally designed cyberattacks more and more ... We could imagine a very dark scenario when the major power plants in this north-east region would be cyber-attacked, and the damage could be really huge. But I can confirm that our public authorities are dealing with that. (Researcher, 2021)

Two government strategy documents relate to cybersecurity: the "Digital Agenda 2020 for Estonia" (2018) and the "Cybersecurity Strategy" (2019). This preparedness is valid, as the risk of cyberattacks appear to be rising. Figure 5.3 summarizes the key aspects of Estonian security and defence policy.

## 5.3 Perceptions of Russia as a Landscape Pressure at the Intersection of Energy and Security

This section examines landscape pressure for the energy transition in Estonia, focusing on the ways in which perceptions of Russia as a landscape pressure for Estonian energy policy have formed. Policy documents from the period 2006–2020 mentioned a range of factors that can be interpreted as landscape pressures (Kivimaa and Sivonen, 2021). In 2006–2010, these included globally growing competition over energy, risks of cyberattacks, and land-use-related conflicts. At that time, the documents pictured Russia as a substantial landscape pressure and a global superpower ready to utilize military force as well as energy as political tools in international relations. Policy document material was limited during 2011–2015, so landscape pressures were more difficult to detect. Uncertainties in global energy markets and risk of attacks against energy and digital systems were mentioned. During 2016–2020, similar pressures as in the first period of analysis were mentioned. However, Russia was mentioned less, and instead the instability of the global economy and international finance were emphasized. Nevertheless, Estonia has quite consistently perceived Russia as a significant landscape factor, shaping the Estonian state's actions in energy and security policies.

Perhaps due to deep historical and cultural issues, alongside the rather short period of independence from Russia, perceptions of Russia as exerting a substantial landscape pressure on the Estonian energy regime are prevalent in the expert interviews. Thirteen out of sixteen interviewees considered Russia to be a considerable security risk in the energy–security nexus both prior to and post 2022.

We see a very determined and aggressive Russia, rising from the ashes threatening Europe, Finland and the Baltic States. (Civil servant, 2020)

There has always been a strong distrust toward Russia, and Russia has demonstrated several times how they might use energy to influence policy. (Researcher, 2021)

The impact of this risk is evident in the plan of the Estonian state to desynchronize its electricity system from Russia and in complicating the oil shale phase-out in Ida-Viru County (see details in Section 5.5). The direct impact of this landscape pressure has been the slowed-down destabilization of the domestic (oil shale-based) part of the fossil fuel regime, as such pressure was largely not supportive of renewable energy niche development before 2022 (see also Sillak and Kanger, 2020).

Although most of the interviewed experts were in consensus, the perceptions of three interviewees differed. Two experts were somewhat more positive about the influence of Russia, although they still recognized certain risks.

A real conflict between Estonia and Russia, or Russia and Ida-Viru County is very unlikely. (Researcher, 2020)

One interviewed politician mentioned that Russia has been a stable energy partner to Estonia and, contrary to others, believed that desynchronizing from the Russian grid might increase instability:

Well on energy side Russia has been a very good neighbor … electricity has been flowing freely all that time and given us this security of the grid. (Politician, 2021)

The perceptions of Russia as a landscape pressure were largely unchanged as a result of the 2022 attack on Ukraine, due to the preexisting negative perceptions. One interviewee mentioned that, after, things were even more openly discussed as a result. Another interviewee noted:

Well, they have lost every bit of trust that we still used to have. I mean we were never really very optimistic, but now we just regard Russia as an aggressor country. (Civil servant, 2023)

The analysis shows that Russia has been perceived as a dominant landscape pressure for Estonian energy policy from the security perspective. This pressure has remained throughout the studied period, from 2006 until 2023.

## 5.4 Policy Coherence and Interplay

Policy coherence and integration were studied by focusing on synergies, conflicts, and administrative integration between the policy domains of energy and security/defence. To do this, analyses were conducted on Estonia's key energy and security policy strategies, published during 2006–2020, and interviews conducted in 2020–2021 and 2022–2023. The results of these analyses have been published in journal articles (Kivimaa, 2022; Kivimaa and Sivonen, 2021) and some of the key findings are summarized in this section.

It is very clear that security from the Russian threat has played a significant role in Estonia's energy policymaking – resulting in a rather unique approach among EU countries prior to 2022, when the focus in the EU was mostly directed toward open and competitive energy markets. A moderate-to-high integration of security was already visible in Estonia's energy policy documents during 2006–2010 – market collaboration with Europe, oil shale, and diversification of supply being the key means to achieve this (Kivimaa and Sivonen, 2021). This integration has become even more visible since 2016.

Kivimaa and Sivonen (2021) noted that energy–security integration has been observable in policy objectives, policy measures, and policy processes alike. However, the objectives have been conflicting from the perspective of decarbonization, emphasizing both oil shale extraction and wind power development. In the policy documents of 2016–2020, security was visible in the objective to redefine security of supply for operational continuity even when transmission capacity between EU member states is lost. Among measures, the energy (de)synchronization project has been vital; the institutional change has been easier to achieve, while technologically the project has not yet been completed. Policy strategy documents have also focused on cybersecurity readiness measures and investments in additional pre-warning systems for defence to enable offshore wind development. This kind of integration is better achieved by involving Ministers of Defence and of Foreign Affairs in the government Climate and Energy Committee, which Estonia has done. The strong security aspect was also very visible in the use of the term "readiness of war" as an important principle for the development of energy systems. Such language was rarely used in other European countries prior to the events of 2022.

Although security from Russia has played a substantial role in Estonia's energy policy, the document analysis showed energy issues were not substantially addressed in the country's defence and security strategy documents; less so than in Finland's, for example (Kivimaa and Sivonen, 2021). This lack of integration may also explain the difficulties and slow progress with offshore wind power expansion.

An analysis of expert interviews from 2020 to 2021 also shows evidence of insufficient formal integration of energy and security policies. Justified on the basis of small country size, it was seen that in areas where policy coherence has been achieved this has been the result of informal connections and bottom-up problem-solving exercises. Strategies, objectives, or shared visions in bringing together energy and security policies have been missing (Kivimaa, 2022). Yet some concrete policy decisions have supported improved coherence between energy and security considerations despite the absence of supportive processes. Importantly, these include the desynchronization process and the decision to invest in new radar technology to allow for the expansion of wind power.

As in many countries, the "National Security Strategy" has framed energy as critical infrastructure and a vital element of national defence. This pursuit is mainly visible in the strategy as ambitions to intensify collaboration with the EU, the Baltic States, and the US, and to keep using oil shale in a "rational" manner – instead of more concrete action proposals. The continued use of oil shale has created a substantial conflict with the progress of the zero-carbon energy transition. Another observed potential conflict was the impact of oil shale phaseout on internal stability in Ida-Viru County. This is explained in more detail in Section 5.5.

Despite the existence and acknowledgment of these conflicts, formal mechanisms and strategies for advancing coherence between the policy domains of energy and of defence and security have been lacking. Policymaking has been reliant on informal communications between the actors in charge. There was a lot of divergence in how actors with expertise on energy and security observed this interaction between the two administrative sectors during 2020–2021. Some agreement that the interaction between key ministries has improved did, however, did exist.

On the positive side,

The Estonian public sector is quite small ... which means that there is no need to create large and heavy administrative systems. We just meet up with the relevant people and have the discussions and try to find a solution which is workable, and which is win–win for everyone. (Civil servant, 2020)

We have very little silos in the Estonian government ... we're well connected and well informed about each other's processes ... each and every year it's getting better and better. (Civil servant, 2020)

And, on the negative side,

Traditionally these two ministries have been held by different political parties. They are not really communicating to my understanding. Probably there are some ways of communication. I would doubt that. (Researcher, 2020)

The ministries still retain quite a substantial degree of latitude in their activities, and we often complain that in this whole comprehensive system of security, there's still too little coordination and, perhaps, an awareness of each other's activities and plans and so on between the ministries ... which sometimes leads to some clashes like, for instance, there's this episode of clash between the wind energy ambitions ... and defence interests. (Researcher, 2021)

While this informal way of working may be a fluid means of policymaking in a country with a population of only 1.4 million people, it allows for a chance element in dealing with potentially pressing issues that tie energy to security and it has reduced the transparency of public administration, as indicated by the interview comments from external actors. As seen with the case of wind power, sustainability transitions may not advance rapidly enough when reliant on informal interactions for integration and coherence. This is also visible in the complications around the phaseout of oil shale; a phaseout that is inevitable if Estonia is to achieve decarbonization of its energy system and the EU emissions target.

A rather significant shift on this front happened following the events of 2022, given that the Estonian state administration had not given much thought to formalizing the energy–security interface before then. A concrete act was that the energy trade with Russia ceased. Another measure departing from previous practice was acquiring the joint LNG terminal with Finland:

This LNG project has been very out-of-the-box thinking, Elering taking an active part in the project due to a government decision. Usually, such projects are left for the private sector. (Business actor, 2023)

Several interviewees remarked that energy issues became politicized after the 2022 crisis, but not securitized in their eyes. It appears that some change in narratives occurred although the risks had already been recognized. Perhaps the most obvious change is cross-political party support for continuing oil shale production, which was a more contentious issue before, while some also reported improved understanding of how energy transition can improve security:

I think that political communication is definitely impacted, because everybody understands that we need them [green energy investments] more. (Business actor, 2023)

Many narratives have changed, what relates to energy security, what relates to the security of supplies, how the country should guarantee its power supplies and gas supplies ... if you think, a year ago, or maybe a bit more, these were not topics at all. (Business actor, 2023)

All the parties support the use of oil shale as a strategic reserve in the future, and [the] majority say that we would need also renewables to come to play, more renewables investments, and we have to remove all the obstacles there. (Business actor, 2023)

More broadly, some experts saw that the shock from the war created positive synergies between energy and security by mobilizing society and increasing communication and interaction between the ministries. The effect, however, seems to have been an increase in volume rather than formalization of previously informal relations:

We are kind of trying to build up the system which is more reliable ... in war situations for example ... These are more on the table and thought over how we can handle them. (Business actor, 2023)

For the past year, it's basically everyday communication and maybe it has come down a bit, but at the beginning all this coordination was just massive. (Civil servant, 2023)

I would say the cooperation is closer, there's more coordination. We kind of try to be proactive, we try to also have better, let's say, communication with potential investors who are interested in having some kind of windmill parks or solar energy parks, so we consult them in early stages. (Civil servant, 2023)

As a result, the administration conducted activities to stimulate the acceleration of renewable energy, such as setting up new, more ambitious goals for renewable energy, speeding up permissions to erect wind turbines, and identifying gaps in legislation potentially hindering renewable energy deployment – contributing to the process of expectation-building in niche development. These measures were influenced by the RePowerEU policy launched in May 2022 by the European Commission in response to the energy crisis:

Everything which is coming from the European side, meaning building the new wind farms or solar parks, must be quicker and must be easier, that we are currently working on, and trying to make everything faster. (Civil servant, 2023)

There are cases where one wind farm, due to the complexity of the permitting, needed two, sometimes even three, environmental impact assessments, so we are trying to make the processes leaner, quicker, and easier in order to get more renewables in. (Civil servant, 2023)

The political pressure has made some quite remarkable steps when it comes to, let's say spatial planning. It used to take a quite long time from the idea to full windmill for example, ... now the process is much ... quicker in a way. (Civil servant, 2023)

At the same time, activities to support the existing fossil fuel regime have continued due to the energy crisis. These include the LNG project, establishing a new natural gas reserve (stockpiling), and slower decommissioning of combustion plants, and these have partially restabilized the regime despite decreasing gas consumption during the energy crisis. The stockpiling was perceived by one interviewee as a shift from market-based thinking to a more security-oriented one.

More broadly, the last few years have seen increased focus on critical infrastructure; in other words, the vulnerability of cross-country underwater energy interconnections on which Estonia is dependent, as well as Russian control over the cooling water used by Estonia's Narva power plant. One expert noted that this risk had been acknowledged previously but the likelihood assessed as negligible – now all risks are reviewed from a different perspective. Section 5.5 illustrates three different cases that highlight how energy and security have interacted within Estonian policymaking.

## 5.5 Niche Development, Regime (De)stabilization, and Positive and Negative Security

In the analysis of expert interviews, three specific cases that relate to niche development and regime destabilization clearly emerged in the energy–security interface in Estonia. The first one deals with the importance of oil shale and the potential economic and geopolitical security concerns that are arising from the phaseout of oil shale in one of Estonia's poorest regions, which also has a significant Russian-speaking population. The second issue connects to the arduous expansion of wind power and how it has been impacted by the operation of the defence regime's air surveillance radars. The third interesting issue is the desynchronization of Baltic countries from Russia's electricity system, where the principal driver has been that of national security, but the move can also be seen to benefit the zero-carbon energy transition. Hence, the three issues relate to the destabilization of the existing energy regime, niche expansion, and regime reconfiguration via the desynchronization project.

An emerging issue related to the expansion of the renewable energy niche is the supply of critical minerals and metals needed for renewable energy technologies. However, this is not discussed here as a specific case, because during the

2020–2021 interviews this aspect was not debated as a security-related challenge in Estonia by the interviewees. The topic did receive attention during the later interviews. For instance, a decision has been made to invest in critical materials-related R&D, while expanding mining activity has its own environmental consequences. Developments have also occurred in the manufacturing side. Silmet, an Estonian subsidiary of the Canadian company Neo Performance Materials (NPM), has announced plans to build a magnet factory and an R&D center in Narva. Therefore, the critical materials question is pertinent for the future development of the energy–security nexus in Estonia. Section 5.5.1 discusses the three cases outlined.

### 5.5.1 Oil Shale Phaseout in Ida-Viru County

Since around 2014, oil shale has contributed up to 85 percent of Estonia's electricity production (Kanger and Sillak, 2020). Except for 2019–2020, this has been a negative development with regard to a zero-carbon energy transition, explained by concerns of national security, economic gains, and energy independence. Estonia has also used its oil shale expertise to advance its position internationally, by developing processing plants in Jordan and the US (Crandall, 2014). Becoming a worldwide expert in oil shale meant that Estonia was recognized by other countries and could make energy alliances. Such alliances led to increasing support for Estonia against the Russia threat, such as the US speaking on behalf of Estonia during its negotiations with NATO in 2004 (Studemeyer, 2019).

EU environmental and climate policy has been a significant driver for oil shale regime destabilization in Estonia. For example, in 2018, Eesti Energia announced it would close four old and inefficient oil shale electricity production units in its Narva power station to comply with the air emissions directive (IEA, 2019). Further pressure for the oil shale industry has been imposed by EU decarbonization targets:

Now for the first time in history Estonia has this serious prospective of phasing out oil shale. And phasing out oil shale is a huge political issue. (Civil servant, 2020)

Most of the Estonian oil shale industry is located in Ida-Viru County, a northeastern region of Estonia at the Russian border. It can be described as a postindustrial area dominated by Soviet industries, where a lot of oil shale industry was developed. After World War II, the region was subject to Moscow's policy to minimize the number of Estonians in the region by forcing the relocation of Estonians elsewhere and locating Russian people to the region, while many Estonians also made their own decision to migrate to economically more attractive parts of the country (Holmberg, 2008). In current times, over 80 percent of the population is estimated to be Russian-speaking and over 40 percent to hold Russian citizenship.

The region has less than 10 percent of Estonia's population, and only 58 percent of working-aged people are employed (Statistics Estonia, 2021). It has the highest unemployment rate coupled with low average income (Prause et al., 2019). It is, thus, the most socially deprived region in Estonia and, consequently, the oil shale industry has been an important source of income. Without this industry, the already poor social conditions could become worse (Holmgren et al., 2019):

Considering that it actually is deindustrialized or postindustrialized, social issues there are pretty extensive. [The] population is poorer than average in Estonia apart from some very rural and forgotten areas in the south of Estonia … We are talking about big cities. In terms of urban population, they are certainly the poorest. (Researcher, 2020)

Thus, the decisions regarding oil shale phaseout need be balanced with minimizing further unemployment and economic hardship, and may create risks of internal instability in Estonia and worsen economic insecurity in the region. The interviewees also pointed out a risk of Russian interference in issues in northeast Estonia due to the large Russian-speaking population; a worry that generated especially strong discussion after Russia invaded Crimea in 2014. This is perhaps connected to the failed attempt at independence from Estonia by the city of Narva in 1993 (cf. Sillak and Kanger, 2020).

The conclusion of this discussion was very strongly that it is extremely unlikely due to several reasons. People's well-being and livelihoods are better on the Estonian side of the border than the Russian side. The local population is well aware of that. And of course, Estonia belongs to the EU and NATO … There have been some attempts by Russia to influence and connect there to the Russian-speaking population, but their influence has been very small. (Researcher, 2020)

You just tell them, okay, we go away from shale oil because it's one of the most energy intensive fuels. We close the factories. We close the mining but then the social tensions will even amplify. They will get amplified. The other side of the Narva river would also react to it and maybe add fuel to the fire. We don't know anything. (Politician, 2021)

Politicians have tried to keep tensions as low as possible with the Russian nationals who, throughout the years, have not integrated well into Estonian society (Crandall, 2014). Effectively, the residents of Ida-Viru County have been found to have the lowest trust of all of Estonia's regions in the president of Estonia, the public governance, and the judicial system (Ministry of Defence, 2021). The 2023 election result showed that the Russian-friendly United Left Party had increased their popularity by receiving 15 percent of the votes in the region. This debate is also intertwined with the rise of far-right populist politics in Estonia, which has attracted an increasing number of votes from Russian speakers, despite EKRE's negative stance toward Russia and the rights of the Russian-speaking population. EKRE is against EU climate policy and, in 2020, proposed the withdrawal of Estonia from the EU emissions-trading scheme. There is, however, consensus between EKRE and the Center Party to save the northeastern part of Estonia from collapsing, despite their differences in politics concerning the integration of the Russian-speaking population into Estonian society. EKRE has used the unemployment and self-determination of Ida-Viru County to support its anti-decarbonization rhetoric (Yazar and Haarstad, 2023). In turn, Kersti Kaljulaid, president of Estonia during 2016–2021, very openly communicated to the population of Ida-Viru County that oil shale is the past and a new plan for the region is needed.

Ida-Viru County and its oil shale phaseout is one of the issues that the EU Just Transition Mechanism is trying to alleviate, with attention paid to employment, the environment, and social inclusion:

So, this is where the just transition mechanism and program comes in. We are in Estonia with different ministries and also with the local municipalities working hard to put together a program which could really then pose as [an] alternative, and which would smoothen the transition which is inevitably happening there if the carbon neutrality goal is achieved, as I'm sure it will be. (Civil servant, 2020)

In 2022, the EU Just Transition Fund awarded almost 19 million euros to launch a magnet factory in Ida-Viru County, which also responds to EU security-of-supply concerns and the dependence on China for magnets (Kalantzakos et al., 2023). The magnets are vital, for instance, in electric vehicles and, therefore, the investment relates both to the economic security of Ida-Viru County and the security of European supply for technical components required in the energy transition (policy synergy).

The aftermath of 2022 saw some revived interest in the use of oil shale. It became competitive against other energy sources when energy prices increased because of the 2022 crisis, and its continuation has received political support since then – indicating a restabilization of the regime. At the same time, however, the goals for decarbonization are pursued, increasing policy incoherence within Estonian energy policy.

We saw sort of a revival of oil shale power generation in Ida-Viru. So, for years before that, the trend in Ida-Viru was that less and less electricity is produced from oil shale, less and less workforce is needed in [the] oil shale industry. Last year, it changed. Last year, those power plants or the energy sector in Ida-Viru recruited, I don't know, more than 1,000 people back to work, because, all of a sudden, those power plants were needed in the regional energy market and also in Estonia. And, of course, from that perspective, it made people in Ida-Viru think, whether the green transition all of a sudden has stopped, and whether the situation we saw last year is a new normality. (Civil servant, 2023)

Eesti Energia is still having the goal in its mind to be 100 percent renewable on electricity by 2030, and, by 2040, [in its] entire [energy] production. So, the plans are still there, the plans what to make from oil shale after we don't produce electricity [from it] anymore. So, the fact that, yeah, now we had to use more, and still we don't know how much more, we had to use oil shale to produce electricity, but the overall vision is still toward neutrality. (Civil servant, 2023)

The restabilization of the oil shale regime has been actively supported by the far-right political party EKRE since 2018 (Yazar and Haarstad, 2023), while other political parties began backing this more for security-of-supply and economic reasons from 2022. An energy business representative stated that the old parts of oil shale power plants in Narva have now received an extension from 2023 until at least 2026, because Eesti Energia is required to guarantee a certain amount of dispatchable production capacity until 2026.

### 5.5.2 Wind Power and Air Surveillance for Defence

Alongside biomass, wind power is one of the key renewable energy sources used in Estonia, with significant plans to expand. It is, therefore, the most promising renewable energy niche for Estonia. The country's geographical positioning is favorable to wind power production (Crandall, 2014). However, offshore wind-power development has had a complicated beginning, with a lack of marine spatial planning regulation prior to 2015 and strong opposition from residents (Tafon et al., 2019). Perhaps a bigger problem – and one that concerns national security more closely – has been the conflicting views of the Ministry for the Economic Affairs and the Ministry of Defence regarding the expansion of wind power. This conflict has been openly acknowledged by those working in the domains of energy and security. During the 2020–2021 expert interviews there were differing perceptions of this conflict between the zero-carbon energy transition and national security. In contrast, by the 2022–2023 interviews a more consensus-based view had formed and solutions had been found via investment into new defence air-surveillance radars.

The conflict was substantially drawn out. In 2006, an offshore wind project off the coast of Hiiumaa, with plans to construct 100–200 turbines with output capacity from 700 MW to 1,100 MW, was halted when the Ministry of Defence suspended the permitting processes for all offshore wind proposals (Tafon et al., 2019). The reasoning at that time related to the need to establish a legal framework for regulating and planning such developments. Another case involved two wind turbines in Ida-Viru County, which had received permission from the Ministry of Defence in 2012 but, when constructed, were higher than initially permitted, so were found to interfere with air surveillance radars and signal intelligence (Koppel, 2022).

The Ministry of Defence faces a technical problem, whereby the increasing height of wind power turbines interferes with the operation of their air surveillance system – vitally important for Estonia: "Wind turbines can affect the ability of radar to detect and track airplanes, resulting in negative impacts on national

## 5.5 Niche Development and Regime (De)stabilization

security capabilities" (IEA, 2019, p. 124). While better siting of wind turbines and upgrading radar technology can offer some solutions, the small land area and the geopolitical location of Estonia have created obstacles – with the result being that planned wind-power projects of over 500 MW have been objected to by the Ministry of Defence (IEA, 2019).

Wind-park dimensions have grown three times since ... the beginning of the 2000s ... and they are starting to impact some of our defence systems. (Civil servant, 2020)

There is the issue with radars. Defence radars which cannot function if some sort of wind farm is built right in front of them. So, this is a real serious issue but by now we have overcome it ... there has been a decision to build a new radar. (Civil servant, 2020)

This debate has created quite a lot of conflict between different parties. Some have argued that the Ministry of Defence has prevented the development to an unreasonable extent, while others have been more understanding:

The Ministry of Defence for a long time didn't want to talk about it. It was like a public secret that we have a radar and we are monitoring what's going on over borders. But now they said it publicly and it's not secret anymore. (Researcher, 2020)

The Ministry of Defence has been, let's say, too offensive and not willing to cooperate. Because I think there are solutions that are able to give that clear view back especially to the east side and have also these wind turbines. On the governmental level, we have agreed to build another radar ... But I think still it needs a bit [more] cooperation between this Ministry of Defence or security or military side and the enterprise side to meet both objectives. (Politician, 2021)

The processes resolving these conflicts have taken a long time, with some even resorting to court dealings, but some solutions have been found. In 2018, the government established a working group between the Ministry of Defence and wind power developers to clarify national security restrictions on wind power projects – yet, the IEA noted in 2019 that the government had not offered "a clearly defined planning and permitting process that will allow developers to ... open a pathway to projects to proceed in a manner that does not excessively impact radar and other national security assets" (IEA, 2019, p. 124). A decision to invest in a new radar, opening up quite a large territory for onshore wind energy in northeast Estonia, was made by the government in 2019.

The process of wind power development has, nevertheless, been extremely slow and reveals a significant conflict between national security and zero-carbon transition policy objectives. Between 2017 and 2021 no new wind turbines were made operational as suitable locations approved by the Ministry of Defence had not been found (Resmonitor EU, 2021). The new radar should become operational in 2024, finally allowing the planned expansion of wind power and removing all height restrictions. The court dealings were also noted by several interviewees in 2023 to have ultimately led to much-improved collaboration and coherence between the ministries:

I would say that for all our offshore or onshore developments, there is quite good cooperation between ministries, starting from the environmental ministry, in terms of our [environmental impact assessments] EIAs, then the planning in our finance ministry, and also from our defense ministry. (Civil servant, 2023)

The 2022 energy crisis created a new legislative process that aimed to shorten the planning process by allowing some steps to happen in parallel and resulted in extra funds to acquire new air surveillance radars too; the latter found "extraordinary" by an interviewee from the Ministry of Defence. Several new projects in Ida-Viru County have been planned to increase wind power capacity.

### 5.5.3 Desynchronization from the Russian Grid

Developments in the reduction of oil shale supply, combined with Estonia being a full member of the Nordic electricity exchange Nord Pool, have focused attention on desynchronizing Estonia's electricity network from Russia. Initially, after regaining independence, the Baltic States were not connected to the Central European grid and had to sign what is called the BRELL (Belarus, Russia, Estonia, Latvia, Lithuania) agreement in 2001, which formalized the countries' synchronous operation with the Russian electricity system (Juozaitis, 2020). A complete synchronization with the Central European electricity network and desynchronization from the Russian power system were announced as mutual strategic priorities by the Baltic States in 2007 (Juozaitis, 2020) and have been in preparation since 2009 (IEA, 2019). A principal reasoning for this has been national security, via breaking ties with Russia. Not everyone in the Baltic States was united on this at first. A common perspective was, however, reached from around 2010:

Fifteen years ago, there was a lot of lack of consensus about synchronization with continental Europe or desynchronizing from Russia, so our thinking was not synchronized between the Baltic States at all. So, there's a very different picture from where we are now. (Researcher, 2021)

The last ten years I would say, all the three Baltic States are in cooperation with Finland, Sweden, and Poland to build a network which could be sustainable and could be autonomous also from Russia's electricity ... Up to 2025, all the three Baltic States, according to the plans, should be ready to desynchronize from Russia's electricity networks. (Researcher, 2021)

In 2014, after Russia stopped the flow of gas to Ukraine, causing an energy crisis, the EU also began to pay more attention to synchronization. Two political roadmaps to advance synchronization were signed by the Baltic States, Poland, and the European Commission in 2018 and 2019 (IEA, 2019). The Baltic Synchronization Project, to improve the security of supply of the Baltics, has been supported by the EU to the tune of over one billion euros and is expected to be completed by 2025. The

Baltics had planned to stop all electricity trade with Russia once the synchronization was fully complete, but the 2022 events led to the electricity trade being halted early, in May 2022. Until desynchronization is completed, however, the Kremlin is argued to have "a very detailed and up-to-date picture on the situation of Lithuanian, Latvian and Estonian power systems" (Juozaitis, 2020, p. 5). In the meantime, potentially strategic information is being passed on to Russia, while the Baltic States have agreed to speed up desynchronization from the timeline initially planned.

Estlink 1, a submarine electricity cable between Finland and Estonia, was opened in late 2006, with national transmission-network operators Fingrid and Elering becoming its owners in 2013; it was followed by Estlink 2 in 2014. Other new connections that have also benefited the Estonian energy system are the LitPol Link between Lithuania and Poland, in 2015, and NordBalt, which began to operate between Sweden and Lithuania in 2016.

Simultaneously, Russia advanced its own desynchronization from the Baltic States. It did so in order to have "a chronological advantage over the Baltic States" and the possibility of unexpectedly withdrawing from the BRELL agreement (Juozaitis, 2020, p. 6). An interviewee noted that Russia had already shown its power position by not supplying emergency electricity to the Baltic States in June 2020, when substantial blackouts occurred following disruptions in the electricity network. This caused a tenfold increase in electricity prices, and even worse blackouts would have occurred if Poland had not supported the Baltic countries with an emergency supply of power.

In 2022, Estonia's electricity market was heavily import-dependent, especially during peak consumption periods (ERR, 2021). For some, this has raised concerns about peak consumption (e.g., cold weather) coinciding with those of the neighboring countries, such as Finland. Thus, diversifying and developing renewable energy is increasingly becoming a priority to reduce energy supply risks (but coexists with demands to continue the oil shale industry). According to the expert interviews, the energy crisis also led to a search for ways to speed up desynchronization, improving the readiness to desynchronize to within twenty hours if necessary (for instance, if Russia interfered with Estonia's electricity supply):

Now the war has really made all of us understand that it needs to be quicker. Russia cannot be relied upon in any aspects of life. So, we would like to speed it up, but it's a completely technical issue. It really, it's mostly about hardware. Meaning that we have to build those, I think in the Baltic countries altogether, nine huge machines which will guarantee this synchronization, frequency synchronization. (Civil servant, 2023)

The desynchronization project provides synergistic developments for improving Estonia security vis-à-vis Russia and advancing the zero-carbon energy transition. Electricity supplied via the Estlinks and the NordBalt is more likely to be produced fossil-free than that supplied from Russia or produced with oil shale.

Further, the desynchronization is connected to policies advancing the use of fossil-free energy sources, with some risks for the security of the energy supply, especially if the expansion of wind energy infrastructure keeps being delayed.

## 5.6 Concluding Remarks

Estonia is an interesting case of how "negative security" interests, that is, threats to national security and geopolitics, tie into energy policy and how this, on the one hand, complicates and slows down the zero-carbon energy transition while, on the other, has also advanced it to a degree by the country disconnecting from Russian fossil fuel supplies. Energy has been, to a degree, securitized since Estonia's independence, and is mainly thought of from the perspective of negative security (security against threats; see Gjørv, 2012). The substantial threat of Russia has been the prevalent landscape pressure on the energy regime, although other landscape pressures have also been noted – such as climate change, cyber risks, and globally increasing energy demand.

Due to the perceived high Russian landscape threat, first, the impact of wind turbine niche expansion on defence radars made the Ministry of Defence cautious about wind power until recently and, second, there has been caution regarding how Russia would react to the phaseout of oil shale located in the bordering Ida-Viru County, which has a large Russian-speaking population. EU just transition efforts in connection to supporting the oil shale phaseout, that is, regime destabilization, and to new industrial activity in Ida-Viru County, have introduced an element of "positive security" into the Estonian energy–security nexus. These efforts have addressed the well-being effects of the energy transition in this relatively poor region. Yet this is contrasted with diverging political interests related to resistance to EU energy policies.

Progress around electricity network integration with Central Europe, which reduces the energy independence of Estonia, resonates with ideas around redefining energy security by accessing cross-border grid communities (cf. Blondeel et al., 2021). Yet this aim coexists with the old security narrative constructed around oil shale and energy independence. Relatively little attention has been paid by policy actors to the security implications of the expansion of the energy "niches," such as solar and wind power, while these implications have clearly increased via newly emerging attention paid to critical materials. More broadly, the extensive efforts already made by Estonia in cybersecurity will also help address the security implications of a new energy system based on renewable energy and the expanding interconnected electricity system. Figure 5.4 summarizes key energy security aspects and their impact on energy transition in Estonia.

The events of 2022 changed the energy–security nexus. Although security was already present in energy policymaking (more so than the presence of energy in

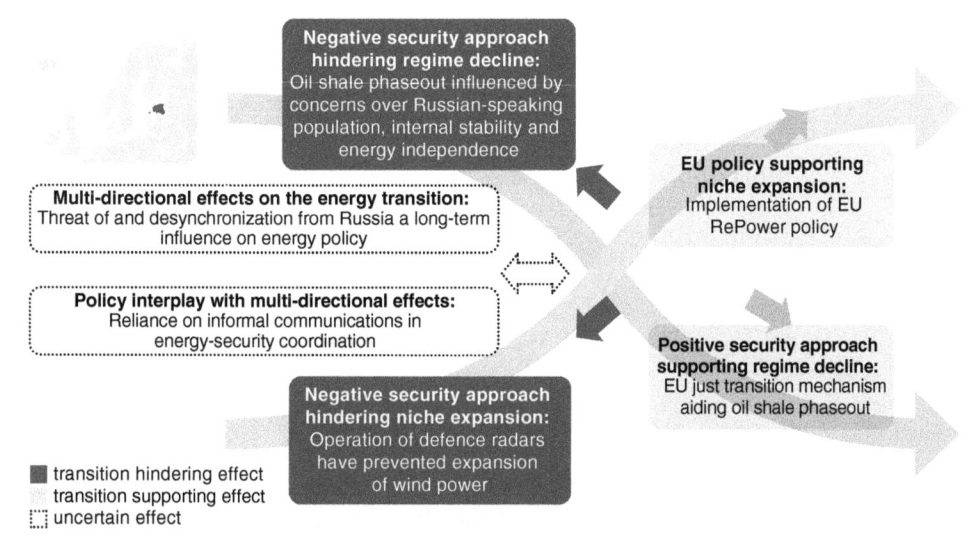

Figure 5.4 Key energy security aspects and their transition impacts in Estonia, 2006–2023.
Source: Kivimaa, Finnish Environment Institute, 2023.

security policymaking), the dialogue became more open after 2022. New solutions for wind power were sought even more actively, with additional radars enabling wind power niche expansion as a measure to improve the security of energy supply. At the same time, however, fossil fuels were restabilized by a new gas reserve, a joint LNG terminal with Finland, and improved political consensus about continuing oil shale production at least for some time. It remains to be seen how these two aims will be balanced in the years to come.

The conflicts situated on the energy–security interface involving oil shale phaseout and the expansion of wind power have had a significant hindering effect on the zero-carbon energy transition in Estonia. These conflicts are unlikely to be resolved immediately due to heightened attention on geopolitical security since Russia's attack on Ukraine in the winter of 2022. However, recent institutional changes have supported the zero-carbon energy transition, most importantly the goal for 100 percent renewable electricity by 2030, set in 2022. Nevertheless, the EU decarbonization targets and Estonian renewable energy niche actors face a counterforce of strong and established energy regime actors and entrenched party politics. The perceptions and established networks of these regime actors may be hard to overcome. Informal governance structures could be employed for flexibility but may also cause a barrier for new entrants and result in a lack of transparency regarding the energy–security nexus.

# 6

# Finland

*Ambivalent Links between Energy and Security*

Finland is a small Northern European country in population terms, with only 5.5 million residents. Yet it covers a rather large geographical area – 338,440 square kilometers. The country is situated between Russia in the east, Sweden in the west, and Norway in the north. Estonia is only around 50 kilometers away, across the Gulf of Finland. In the past, Finland has been both a part of Sweden and of Russia, with independence gained in 1917. Both historical connections have partly influenced energy policymaking in Finland, and the country has had active energy trade across eastern and western borders, with the former halted after Russia attacked Ukraine in 2022.

Since World War II, Finland has gradually transformed from an agricultural society to a technological one, where the development of the forest and telecommunications industries was particularly significant. The forest industry has had a profound influence on both the demand for energy and the use of forests. Early on, the industry became an energy producer because postwar industrialization raised energy consumption. Hence, forest industry companies invested in electricity production from hydropower and established their own energy company, Pohjolan Voima (PVO), in 1943. In the 1970s, PVO expanded with condensation and nuclear power plant investments. Innovation in pulp and paper production processes enabled forest industry companies to produce bioenergy via their byproducts from the 1980s and 1990s onward, improving the energy economy of the forest industry. After recession in the early 1990s, the telecommunications industry acted as a spearhead for innovation policy. This industry experienced somewhat of a decline after Microsoft purchased global telecoms leader Nokia in 2014 and, later, ended most of its operations in Finland.

Finland is a rather interesting case to study energy–security relations for multiple reasons. First, its domestically available fuels are limited. Besides wood fuels, Finland uses peat in energy production. Peat provided 7 percent of total energy consumption in its peak year in 2007 but, in 2021, was reduced to 2.7 percent of

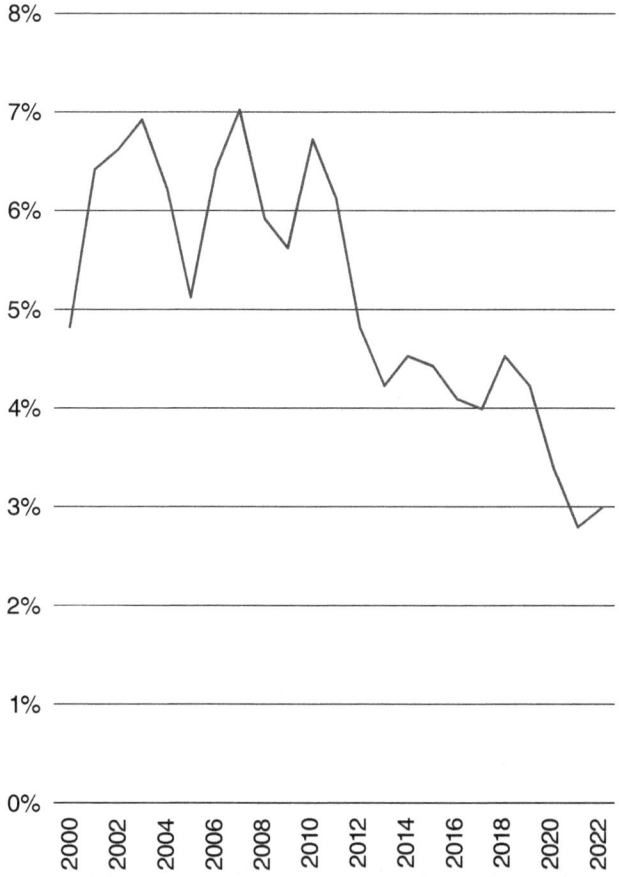

Figure 6.1 Percentage shares of peat in Finland's total energy consumption, 2000–2022.
Source: Statistics Finland (2023).

total energy consumption (Statistics Finland, 2023), with a policy goal to halve the use of peat from the 2020 level by 2030 (Figure 6.1). During the 1993–2012 period, peat provided a minimum of 5 percent of Finland's energy consumption. A strong peat lobby has existed since the 2000s, utilizing security as an argument when favorable. One campaign occurred after the 2006 gas disruptions between Russia and Ukraine, resonating with fears and concerns about energy availability and price (Lempinen, 2019). Following the events of 2022, policies for the peat phaseout were temporarily relaxed and an emergency stockpile of peat created. This case is explored further later in this chapter.

Second, Finland is a neighboring country to Russia, with a 1,340-kilometer shared border, which has meant defence preparedness since Finland's independence in 1917. However, the country also experienced a culture of "Finlandization" after World War II, described as the "adaptive acquiescence to the will of the Kremlin during the Cold War" (Arter, 2000, p. 688). This was followed by the search for positive business relations with Russia and with the avoidance of negative remarks toward Russia in energy policy, and more generally in public discussion. Even after the EU imposed sanctions on Russia after the annexation of Crimea in 2014, Finland kept shoring up its relations to Russian energy value chains and maintained the framing of Russian energy (merely) as an economic topic (Höysniemi, 2022). At the same time, Finland has maintained strong territorial defences in the post-Cold War era due to its historical experiences with Russia (Pesu and Iso-Markku, 2020).

Third, in 2019, the coalition government led by the Social Democrats set an ambitious climate policy goal of a carbon-neutral society by 2035, while the previous government had decided to ban the use of coal in energy generation by 2029. These moves have been supported by the already declining trend of the share of fossil fuels in Finland's energy production from about 2010 and an acceleration of new renewable energy sources, especially wind power, from about 2014. During the initial study period 2006–2021, prior to the European energy crisis, the share of oil decreased from 25 percent to 19 percent, coal from 14 percent to 6 percent, and natural gas from 11 percent to 5 percent (Statistics Finland, 2023). While the overall share of natural gas was low in 2022, this was still problematic for Finland because all the natural gas had been imported from Russia. Finland managed to secure a liquefied natural gas (LNG) vessel to compensate for some gas supplies. This is the setting in which the analysis of security and energy transitions takes place.

This chapter presents the country case study of Finland. It first describes Finland's energy and security regimes. It then continues with the analytical sections, drawing on Chapter 4: namely, the perceptions of Russia as a landscape for energy transitions; policy coherence and interplay between energy and security regimes including the level of securitization; and, finally, positive and negative security related to niche development and regime (de)stabilization. The project this book is based on studied Finland's energy and security-related government strategies published since 2006 and conducted two rounds of interviews with energy and security experts, the first between September 2020 and April 2021, and then the second between December 2022 and January 2023. This chapter draws on these materials and on related literature and selected policy reports.

## 6.1 Energy Regime

The energy sources of the Finnish energy regime are based on a mix: (relatively limited) domestic fuels (wood-based fuels and peat) and imports of coal, oil, and natural gas (which have been significant since World War II). Therefore, wood has been an important energy source directly and via forest industry byproducts. The decarbonization of the electricity sector has already reduced dependence on these energy forms but transport – and to a lesser degree heating – still rely on imported oil and natural gas. Overall, the energy profile has been variable with no single source dominating.

Finland has a relatively fixed amount of hydropower as an important balancing capacity, with its further construction restricted since the 1980s for nature protection reasons and because the largest rivers were already utilized for hydropower in the 1960s. While, in the 1950s, over 90 percent of Finland's electricity was produced by hydropower (Kivimaa, 2008), hydroelectric plants began to face opposition from local communities due to disrupted fishing and farming activities, leading to increased support for nuclear power in the 1960s and 1970s (Myllyntaus, 1991). Finland, therefore, heavily relies on hydropower-based electricity imports from Sweden and Norway.

Nuclear power has been part of the mix since the late 1970s, spurred by the worldwide oil crises of that time. In the 1970s and 1980s, four nuclear reactors were constructed. Half a century later, in 2022, the fifth nuclear reactor, Olkiluoto 3, began operating, but suffered from technical difficulties, limiting production, so only began full operation in 2023, about fifteen years behind the initial schedule. Attitudes to nuclear power have varied over time. The pronuclear group has aimed to depoliticize nuclear power with argumentation about technical safety, while the antinuclear camp has aimed to make the issue more political (Ylönen et al., 2017), raising nontechnical questions related to nuclear power.

Biofuels were important in the early 2000s, especially in the form of black liquor, a byproduct of the forest industry, and later also via the direct use of wood for energy. Besides this, various forms of biofuels and technologies have characterized the development of bioenergy (Kivimaa, 2008), alongside a battle for the use of wood for different purposes. Bioenergy was initially perceived as "the fossil-free source" for Finland, but it became more contentious when wood energy's real impacts on greenhouse gas emissions and on carbon sinks became more widely considered in the EU. Increasing concern has been placed on the reducing carbon sink of the forests when the use of wood has increased.

The year 2013 was described as the time when the wind power niche began taking off in Finland (Haukkala, 2018). Wind power has developed rapidly since 2014, contributing almost 12,000 gigawatt hours and 14 percent of electricity consumption in 2022 (Finland's Wind Power Association, 2023). The capacity in 2022 was circa 5,677 megawatts (MW), with further 44,000 MW land-based and 10,000 MW offshore wind power structures planned (Finland's Wind Power Association, 2022).

Another dominant feature of the Finnish energy regime is the district heating system initiated in the 1950s that covers about a half of Finland's residential and service buildings, and is closer to 90 percent in cities (Schönach, 2021). However, this is partly being replaced by an expansion of ground-source and air-source heat pumps as more sustainable heating sources than hydrocarbon and biomass-powered district heating. The expansion is taking place especially in detached housing stock, while for larger buildings district heating still dominates. The district heating system is seen as an important means of energy storage, especially considering the shifts currently taking place toward the advancement of electrification.

Fortum, a partially government-owned company, is the largest provider of heat and power. It had heavily expanded into Russia prior to 2022 and it owned part of the German Uniper corporation before Uniper's economic difficulties in the same year and until the German government bought Fortum's share of Uniper. The second-largest energy company is Helsinki Energy, operating in the capital region and owned by the city of Helsinki. The manufacturing industry-owned PVO and the Swedish energy company Vattenfall are the third- and fourth-largest energy sellers.

The governance of the Finnish energy regime has been perceived as rather consensus-seeking and stable, with a small number of the (largely nonparty political) energy elite in power (Kainiemi et al., 2020; Ruostetsaari, 2010). There are close connections between certain economic interest groups and public authorities, where these groups can influence policymaking, for instance, regarding energy and the natural environment (from here onward "the environment") (Vesa et al., 2020). Interestingly, citizens have high trust in experts as decision-makers and are less keen for politicians to be in charge (Ruostetsaari, 2017). This has made it easier to depoliticize energy-related issues. The energy elite that consists of influential actors has, in turn, given more weight to economic competitiveness than climate change, the environment, or security (Ruostetsaari, 2017; Vesa et al., 2020). Nonetheless, in 2009, strategic policymaking assumed that fossil fuels will remain the most important energy sources in the coming decades (PMO, 2009), while, in the same year, the Energy Industries Federation published a vision for carbon-neutral energy. Many incumbent energy actors neither pursued the energy transition actively nor perceived Russia as a security threat (Höysniemi, 2022; Kainiemi et al., 2020). Therefore, the events of 2022 functioned as a major "landscape shock" to the established energy governance system, requiring a new kind of perspective that also accounted for geopolitical security.

As a result of the established energy governance system in Finland, Russia maintained its position as a major energy exporter to the country until 2022. During the 2006–2011 period, 12–14 percent of electricity was imported from Russia, but experienced a major decline in 2012, when capacity payments introduced in Russia made export less profitable. Nevertheless, the share of electricity imports experienced a gradual increase subsequently, apart from 2020, when electricity was very inexpensive in the Nordic countries (see Figure 6.2). For other energy sources,

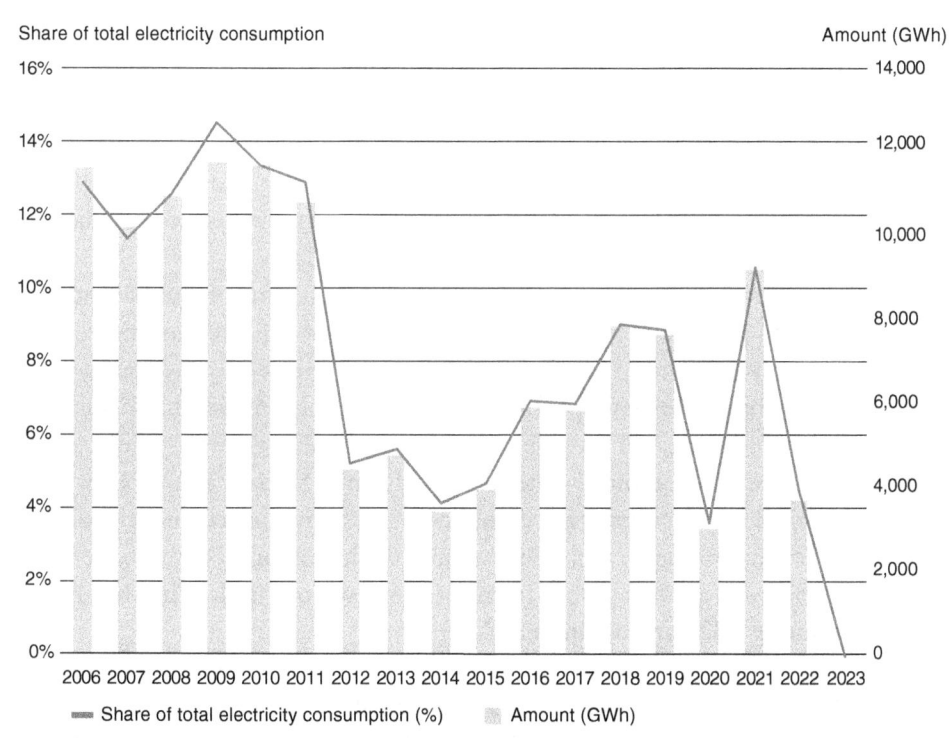

Figure 6.2 The amount of electricity import from Russia to Finland and its share of total electricity consumption, 2006–2023.
Source: Statistics Finland (2023).

the import dependence has been higher – in total 34 percent in 2021 – the largest imported source being oil, followed by nuclear power, natural gas, wood fuels, and coal (Statistics Finland, 2022). A gas pipeline was built from Russia to Finland in the 1970s and 100 percent of natural gas used in Finland came from Russia before 2022. In addition, the oil refinery of Neste Oil specialized in refining Russian oil. Replacing the imports following 2022 was possible, but the prices of energy commodities rose very sharply (Oesch, 2022), experiencing up to tenfold increases in consumer prices.

Energy governance is diffused between different ministries. The Ministry of Economic Affairs and Employment (MEE) has an energy department that coordinates Finnish climate and energy policy and is responsible for energy production and fuel-related issues – the latter together with the ministries of transport and communications (MTC) and of agriculture and forestry (MAF). The Ministry of the Environment oversees building energy efficiency and use, and international climate change negotiations, while the MTC is in charge of transport energy efficiency and use. This can be described as a fragmented energy governance setting.

The National Emergency Supply Authority (NESA) is an essential actor for energy security and other sectors' security of supply, its budget being based on security of supply payments from companies. Its tasks include the coordination of preparedness cooperation between public and private sectors, overseeing arrangements related to national emergency stockpiles, ensuring the functionality of essential technical systems, safeguarding critical goods and service production, and monitoring international developments (NESA, 2023).

NESA reports that the Energy Sector Pools, consisting of public and private actors who prepare for emergencies and exceptional circumstances, have recently been reorganized to match with the changes created by the energy transition. For example, the new Heating Pool is building preparedness skills for nonfossil-based heating and the role of the Power Pool is growing as electricity is increasingly used to replace other energy forms (NESA, 2022). The pools are one example of the collaborative public–private governance culture adopted in Finland for many issues. Another example of this culture is the cooperation between public and private actors from different sectors in drafting fossil-free roadmaps during 2021 and again in 2024.

Since about 2013, new actors, such as the Clean Energy Association and the Climate Leadership Coalition, have challenged the established energy governance system, alongside the Green Party and environmental nongovernmental organizations (NGOs) (Haukkala, 2018; Kainiemi et al., 2020). Attempts to make the energy transition more visible in party political and government program agendas included, during the 2014–2015 preparliamentary election period, a push from temporary actors, such as the Professor Group on Energy Policy, the Energy Renovation Group (Kainiemi et al., 2020), and the New Energy Policy Initiative (Haukkala, 2018). The pursuits of these new actors have also been supported by some governance changes. For instance, the Finnish Climate Change Panel, composed of fifteen academics selected for one four-year term at a time, has become an important body that advises the government on climate and energy matters and responds to climate change-related consultations regarding new strategies and policy proposals. These have led, importantly, to a decision to phase out coal power together with speeding up favorable policies for renewable energy. Despite the advances made, very little disruptive institutional work has taken place (Kainiemi et al., 2020). Despite the coal phaseout plan, the prolonged debate about the fate of peat – tying in conflicting interests and concerns about employment, energy security, and the environment – is not over (Lempinen, 2019). Moreover, there were calls to establish a new science advice panel on forests and the bioeconomy to create pushback to environmentally oriented preexisting science panels the Climate Change Panel and the Nature Panel; ironically, the new science panel has given similar advice to the older ones.

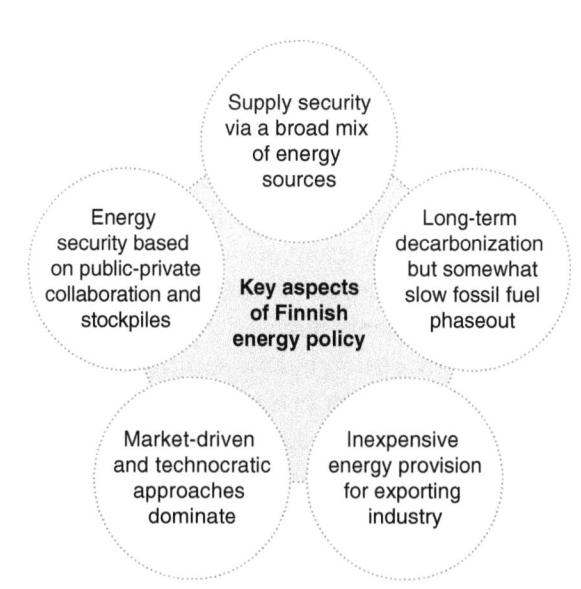

Figure 6.3  Key aspects of Finnish energy policy.

One of the key obstacles to both the energy transition and the broader consideration of security in Finnish energy policy has been the energy policy department of the MEE. Multiple studies have described it being stuck in an economic mindset favoring incumbent forms of energy production. For instance, Höysniemi (2022) describes market-driven and technocratic governance, while Kivimaa (2022) outlines a dominant economic perspective with little attention paid to security and the environment. Further, the ministry has been unwilling to disclose the assumptions in its "Climate and Energy Strategy" scenarios to outsiders (Kainiemi et al., 2020). Overall, the energy department has been slow to change and has lacked an innovation-oriented approach, despite being placed in the same ministry as innovation policy. Some experts, however, argued that the key problem derives from the political level, which has resulted in the fragmented energy governance framework:

When political decision-making is "limping" so does the civil servant machinery alongside it ... If we don't have political willpower, strategy, process, and timetable constructed so that it reaches over several government terms, as in energy policy must be, the civil servant machinery cannot realize things ... We have a minister who has other things in mind first. (Business actor, 2023)

The same issue of political decision-making influencing how the public administration is organized and coordinated has affected the policy interplay between energy and security governance. I explore this in Section 6.4. Figure 6.3 summarizes the key aspects of Finnish energy policy.

## 6.2 Security Regime

After the end of the Cold War, Finnish defence policy changed and the first ever "Security Policy Report" by the government was published in 1995. The traditional components of Finnish defence policy are territorial defence and the Defence Forces, which have gained strong support across different political parties, in essence keeping the possibility of war part of the strategic culture of Finland (Pesu, 2017). It, therefore, has been quite peculiar that this strategic culture has played hardly any role in government energy policy.

Over time, there has been variation in how security and defence policies have been addressed either in combination or separately. Pesu (2017) argued that to divide the government's security and defence policy into two separate reports during 2016–2017 was an important decision that returned the defence administration back some of its past attention given to it in political decision-making. While the key focus of Finland's defence and security policy has been the threat of the East, before the end of the Cold War it was not acceptable to publicly talk about the threat of the Soviet Union, which hindered security discussion altogether. Even after the Cold War, Russia was the most important nation for bilateral trade relations (Nokkala, 2014).

During 2003–2010, Finland gradually shifted from a total defence approach, meaning military and civil defence, to a broader conceptualization of comprehensive security (Berzina, 2020), tying in both negative and positive security approaches. Comprehensive security means an operational model that relies on safeguarding vital societal functions in collaboration with authorities, businesses, NGOs, and citizens, covering issues such as defence capability, internal security, security of supply, functional capacity, and psychological resilience (Finnish Security Committee, 2017). However, it has meant that it is difficult to define the exact boundaries of Finnish security policy. While threat scenarios have become broader, the role of Russia has remained the same after the Cold War (Nokkala, 2014). Energy has been mentioned as part of comprehensive security, but the experts interviewed perceived considerations of energy to be on a rather general level. One expert speculated that energy was not, prior to 2022, made a specific component of security policy, perhaps because this would extend discussions to Russia as a security threat directly connected to questions of energy policy – which was avoided at the time. So, energy was kept depoliticized and desecuritized.

The security regime in Finland comprises the defence administration, foreign affairs, and internal security (e.g., the police force), with links to preparatory activities by the National Emergency Supply Organization, comprising NESA and a broader network of companies. The Ministry of Defence, the Ministry of Foreign Affairs, and the Ministry of the Interior are important governing actors for security. The Defence Forces operate under the Ministry of Defence. The Ministerial Committee on Foreign and Security Policy is one of four statutory ministerial committees and meets with the President of the Republic to prepare important parts of Finland's foreign and security policy. Defence spending has, since 2010, ranged from 1.22 percent of GDP in 2017 and 2018 to 1.96 percent in 2022 (MoD, 2023).

Since about 2010, the defence administration has had to adopt responsibility for new areas. These include, for instance, cybersecurity and climate change. The first "Cybersecurity Strategy" was published in 2013. Cybersecurity issues are located administratively in different places but coordinated by the National Cyber Security Centre placed in the Transport and Communications Agency Traficom. From 2019, cybersecurity in Finland has relied on the national "Cyber Security Strategy" from 2019, international collaboration, and guidance given to companies and other organizations.

The first "Climate Program of the Defence Forces" was published in 2014, initially mostly addressing the energy consumption and efficiency of the defence premises. The third "Climate and Energy Program of the Defence Forces 2022–2025" broadened the focus. It highlighted that the Defence Forces must prepare for the societal transition and recognized that their operation cannot remain solely dependent on fossil fuels in the long term (Finnish Defence Forces, 2022).

Besides the formal security and defence authorities, the National Defence Courses have been running since the 1960s, inviting important members of society to participate. It is a Finnish particularity that aims to improve collaboration between different societal sectors during exceptional circumstances and advance networking between actors operating on comprehensive security.

Overall, there were relatively few changes to the security regime during the study period from 2006 onward, until Finland joined the North Atlantic Treaty Organization (NATO) in 2023. The Finnish security regime has been well prepared for the geopolitical risks from Russia too. What has been lacking is its sufficient coordination with the energy regime (see Section 6.4) and perhaps quite slow awakening to questions of climate change and climate security. Figure 6.4 shows the key aspects of defence and security policy in Finland.

## 6.3 Perceptions of Russia as a Landscape Pressure at the Intersection of Energy and Security

Here, I explore landscape pressures, especially how perceptions of Russia by energy and security actors have formed landscape pressure on Finland's energy policy and how these perceptions have changed over time. During 2006–2010,

Russia is Finland's neighbour, and its democratic development and stability are important. Finland aims to maintain stable and well-functioning relations with Russia. In addition to economic cooperation, collaboration in [the] Arctic and climate questions, for example, remains important. Finland's energy cooperation with Russia is broad and must be interconnected with the development of the EU's Energy Union. Regional and cross-border cooperation with Russia in northern Europe continues at the practical level, which is in the interests of Finland. It is important to support the civil society and direct contact between citizens. In the changed environment Finland must be able to carefully evaluate Russia's development. This calls for more versatile and in-depth knowledge of Russia. (PMO, 2016, pp. 22–23)

Therefore, while there has been implicit deep-rooted caution in the Finnish worldviews, at the same time the official communication was focused on economic relations with Russia, with much less vocality about the geopolitical threats the country poses.

Since around 2010, Russia has become somewhat more distant to Finland while Finland has become closer to the EU. An interviewee described a generational shift in politics that sees Finland primarily as an EU member state that no longer thinks that Russian interests need to be considered first when making decisions. Some interviewed experts also demonstrated increasing awareness of China, the significant dependency on China in many sectors, and China's future influence.

The exceptional events of 2022, when Russia began a war in Europe, had a clear influence on the perception and construction of landscape threats in government energy policymaking in Finland. Abandoning Russian energy was one heading in the new Finnish "Climate and Energy Strategy" from 2022, and this included references to the decisions by the Council of Europe to make Europe independent from Russian imports of gas, oil, and coal as soon as possible. This would have been unheard of prior to the war in Ukraine.

Unlike before, the expert perceptions were uniform post-2022. There was an expression of lost faith and trust in Russia and no change for the better expected in a long time. Some, however, perceived that also a look back is needed to reflect on past policies:

It is important to go through early 2000s re-Finlandization, because we had an operational culture where it was thought that energy trade and interaction with Russia increases Finnish companies' business opportunities in Russia and keeps Putin more benign to us. This politics has suffered a real "shipwreck" and caused large costs also to companies. (Politician, 2023)

The clarity about Russia post-2022 has also led to more ambitious energy transition policies in Finland and a more open recognition of the geopolitical risks associated with different energy policy decisions. Next, I explore the interplay of energy policy with security and defence policies.

## 6.4 Policy Coherence and Interplay

The landscape perceptions about Russia and the dissonance between experiencing Russia as a security threat and a collaborative trade partner before 2022 explain much of the policy incoherence between energy (transition) and security policies that became evident in this study. I explored policy integration, synergies, and conflicts, as well as administrative interaction between the two policy domains by analyzing key energy and security policy strategies published during 2006–2020 and interviews conducted in 2020–2021 and 2022–2023. The results of these analyses have been published in journal articles (Kivimaa, 2022; Kivimaa and Sivonen, 2021) and I outline some of the key findings in this section.

The analysis of the Finnish government's energy and security policy strategies showed moderate integration of energy issues into security and defence policy strategies, whereas the integration of security into energy policy strategies ranged from moderate to low. The low level of integration was observed especially in the concrete policy instruments that the policy documents outlined, and in the way the broader security considerations of the energy transition were not addressed (Kivimaa and Sivonen, 2021).

During 2006–2010, the reference objects at the energy–security nexus of the policy strategies were correlated with "vital functions" of the society that required securing electricity transmission, distribution, and power supply, alongside broader energy availability and supply. However, the referent objects were not only limited to the energy system. Security strategies also mentioned people's health and well-being being affected by disruptions in energy supply and heating networks. In addition, the environment, business activities, and the defence system were recognized as important reference objects related to energy and security. The second period, 2011–2015, was less specific about reference objects. It mostly mentioned the critical energy infrastructure to be secured. Individual mentions of vital societal functions, the environment, and nuclear safety were made. There was no substantial change during the 2016–2020 period: Energy supply in terms of power and fuels was still emphasized as the object of security. There were selected remarks on nuclear safety, data systems, the environment, and the well-being of the population.

The general finding from the interview analysis was that several coordinating elements that have potentially advanced coherence between energy and security policies existed, although instruments or actors that considered both energy and security *in equal measure* were lacking. Examples of elements with potential to advance coherence included the "National Security Strategy," taking a comprehensive security approach; the National Security Council, including energy representation; NESA; the public–private network Power Pool coordinated by NESA;

and the National Defence Courses. However, neither the "Security Strategy" nor the National Defence Courses have typically addressed the energy sector or energy transitions in detail. The Power Pool system, in contrast, was perceived as very effective in securing the energy system in crises – perhaps proven during the 2022 European energy crisis (Kivimaa, 2022).

NESA, in turn, has received critique from the interviewees. For instance, it was claimed to retain "old world thinking" in terms of energy security. Also, elsewhere it has been described as a rather traditional organization, slow to follow contemporary trends and with its broader network comprising incumbent companies (Höysniemi, 2022). Yet NESA has also begun to orient toward the energy transition by publishing its "Energy 2030 Program" in 2019, which rethinks security of supply. In 2022, the Finnish government announced work to update NESA's mandate to react to the requirements of the energy transition.

That organization has got a lot of movement and forward-looking exercises. I am sure this crisis has sped it up. (Business actor, 2022)

NESA has been in the dark for the last fifteen years ... They have little by little managed to build certain preparedness ... now they use a lot of money on cybersecurity and supporting the digital worlds. A lot of good development has happened. (Business actor, 2023)

Commonly, the interviewees believed that interaction between the ministries and agencies responsible for energy and security has not been sufficient. Prior to 2022, the pursuits for coherence mainly incurred informally, as knowledge exchange in energy- or security-connected meetings. Criticism was also directed toward a siloed and fragmented interface where nobody had general responsibility for the energy–security nexus. More recent interviews have also highlighted the lack of high-level coordination within energy policy itself (see Section 6.1). The critique was partly connected to a political culture of not discussing geopolitical security with respect to energy policy, a "business-oriented" style in energy policy concerning Russia, and a lack of security expertise in the MEE department of energy.

A politician explained the observed incoherence in October 2020 as follows:

Maybe there has been unwillingness to mix economic interests and these kind of security interests with each other because, if one needs to build or starts to build a connection, then there is easily a need to make choices.

The historical-cultural legacy of "Finlandization" was also referred to by some as an explanation. The incoherence between energy and security policies implies that the security implications of the energy transition have not received appropriate foresight analyses. Yet, more recently for instance, the geopolitics of renewable energy has received increasing focus. Finland's NATO membership may also increase collaboration at the energy–security nexus, as NATO is developing strategic awareness of this (Bocse, 2020).

The defence administration, and especially the Defence Forces, have provided rather good examples of energy policy integration – albeit on a very pragmatic level. The Defence Forces' Climate and Energy Program has led to improvements in the energy efficiency of their operations but has also addressed climate adaptation to some degree. While little research has been conducted on alternative energy sources, there has been some follow-up activity on the technological development of this front. To better serve policy coherence, these activities need to be scaled up but are also dependent on how energy governance actors respond.

Perhaps the most visible example of synergies is realized in the context of emergency and crisis preparation of the energy sector. The above-mentioned networks, in the form of energy pools coordinated by NESA and associated practices, are likely to benefit security.

What became clear in the expert interviews was that the root cause of incoherence between energy and security governance was the lack of interest of most political parties, with explicit "hints" even given to civil servants not to explore this connection further (Kivimaa, 2022). This perhaps relates to the need that, if Russian-related risks had been better acknowledged in energy policy, Finland would have had to make some hard choices. In other words, improving the coherence between energy and security policies requires changing the objective setting and measures in one or both of the domains. The connection of the energy transition to national security was even less discussed. One business actor noted in 2020:

On the political side it [the security effects of the decarbonizing energy system] interests no one. I can immediately say that it really is not sufficiently, I mean the system-level change, it is not noted, and it may be a slightly difficult issue for political decision-makers.

During the study period, from 2006 to 2023, two clear turning points were relevant for the interplay of energy and security policies. The first one was in 2014, with the Russian annexation of Crimea, which inspired some caution in energy policymaking. Yet, at the same time, it did not create a big enough landscape shock to fix any observed incoherencies between energy and security policies.

Nevertheless, relatively important policy decisions took place in 2014; for example, the publication of the "Energy and Climate Roadmap 2020," a report by the Parliamentary Committee on Energy and Climate Change. Interestingly, the word "Russia" appears only once in this seventy-three-page report, despite Russia's significant role as an energy importer to Finland during that time. Moreover, (energy) security was not used as a term, but security of supply and self-sufficiency were mentioned about twenty times each. Therefore, although there was focus on security it was only in somewhat narrow terms.

The 2014 "Energy and Climate Strategy" was followed by parliamentary approval of the decision in principle to grant permission to build Fennovoima's Hanhikivi nuclear power plant – with one third ownership by the Russian state-owned nuclear power company Rosatom – in December 2014, ten months after Crimea. The decision in principle used security mainly in relation to radiation and nuclear security, not in geopolitical terms. Crimea was mentioned once in relation to the consultation statement by the Ministry of Foreign Affairs that detailed, for instance, that the project would create negative publicity for Finland in Europe and is counter to EU aims to reduce energy dependence on Russia as a reaction to Russia's illegal annexation of Crimea (Formin, 2016). Security of electricity supply was used as one of the key justifications for the parliamentary approval. An interviewee remarked that one explanation may be the historically rather good relations that Finland had with Russia as a supplier of raw materials; although this has also been a dependence by Finland that the Soviet Union and Russia have been able to exploit in negotiations. Parliamentary approval of this nuclear power development did not, however, mean that critical voices did not exist at the time:

In the discussion about Fennovoima, of course Rosatom's ties to Russian nuclear weapons industry were obvious and clear. That is why critique was focused against it by actors who normally support nuclear power. (Politician, 2020)

I was very critical about nuclear power but especially about this provider that is bound to Russian interests quite strongly, because Rosatom is a heavily strategic actor of the Russian government that does not operate on a commercial basis but on a political basis. (Politician, 2020)

In 2014, there was also rather extensive discussion about energy independence (aiming to produce all required energy in Finland) versus security of supply (making sure that there are sufficient energy imports available to complement domestic supply). The latter was accepted in policy due to much higher costs of energy independence and the market-oriented approach of the energy department. The broader aim has, nevertheless, been to reduce import dependency on Russia:

Finland's energy policy has been … the one long policy line [that] has all the time been, and still is, the reduction of import dependence and it has been directed to getting rid of import dependence from Russia. (Business actor, 2020)

A more significant turning point occurred in 2022, when Russia attacked Ukraine. The following higher energy prices and fear of energy shortages when Russian energy flows were lowered, if not completely stopped. In Finland, this created a large shift in energy policy and the political discourse:

If one thinks about a general strategic level, it has been very strong – this kind of burst of the collective bubble … while discussion was actively quietened before … now it is a completely different kind of situation. (Civil servant, 2022)

It also led to some rather extraordinary policy decisions, although not constituting securitization as such. The energy imports from Russia were not directly halted for contractual reasons but the placement of sanctions by the EU, on the one hand, and the demands of the Russian government to pay Russian companies in Russian currency (rubles), on the other followed. The policy decisions deviating from regular energy policy made rather a long list, not all pertaining to security, but which, for instance, included alleviating the high electricity prices paid by consumers. One of the most exceptional policies brought into use, according to the interviewees, was the construction of an LNG terminal and acquiring an LNG ship from Texas to secure gas supply for the following winter: "On 20 May 2022, Gasgrid Finland Oy and Excelerate Energy, Inc., from the United States, signed a ten-year lease agreement for the floating LNG terminal vessel Exemplar" (MEE, 2022, p. 45). Another exceptional policy has been the clear strategic line taken to also discontinue Russian energy imports in the long term. The "Climate and Energy Strategy" from 2022 highlighted the importance of replacing gas as an industrial power source, the launch of an energy-saving campaign to respond to the sanctions and bans placed on Russian gas, and the establishment of an emergency stockpile of peat by NESA. It is clear that some short-term measures are counter to the zero-carbon energy transition, even though the discourse still emphasized the benefits of the transition to energy security.

In essence, the interviews conducted in 2022–2023 illustrated a significant shift in the interplay between energy and security policies in Finland:

Yes, energy has become a strong part of foreign, security, and defence policy, but it has not yet been implemented in our plans as it should. (Business actor, 2023)

Since early 2022, in energy policymaking there has also been more receptiveness to, or even a call for, security messages from the administration and the business sectors alike. In addition, a new security assessment of geopolitical risks was conducted on the Fennovoima Hanhikivi nuclear power plant. The explosions of the Nord Stream gas pipelines also brought into play the need to physically protect the critical energy infrastructure; something which was considered a much lower risk two years previously. While this does not amount to securitization as defined in Chapter 2, it can perhaps be described as a form of partial securitization of previously economic- and decarbonization-oriented energy policy.

## 6.5 Niche Development, Regime (De)stabilization, and Positive and Negative Security

In this section, I explore, via selected cases, the ways in which energy and security intertwine with niche development and regime destabilization (or the lack thereof). Niche development is principally addressed via the expansion of wind power, this being the most extensive new renewable energy source in Finland with the greatest potential. It is also directly connected to national security via tall wind turbines interfering with air surveillance radars and, less directly, via new dependencies on critical materials. While there are other emerging technologies, such as green hydrogen, they are in such an early stage of development that assessing security based on my data sources was not possible. The other cases relate to the stability of the established energy regime. They involve, first, the case of peat and its perceived importance for Finnish energy security, with security arguments playing a role in slowing down the energy transition; and, second, the lack of any security discussion by the Finnish government around the construction of the Nord Stream 2 gas pipeline between Russia and Germany. These cases essentially represent the path dependencies of Finnish energy policy (peat) and the history of "Finlandization" (Nord Stream 2).

Overall, the reconfiguration of the energy regime via electrification – which is neither directly niche development nor regime destabilization – has improved security in Finland. One politician noted, in 2020, that the Nordic electricity market has substantially improved the security political situation in Finland by reducing dependence on Russian energy sources compared to 2006, and that investments in cross-country electricity lines to Sweden and Estonia have also played a role. Generally, the expansion of electrification in Finland (and Estonia) can be seen as a transition that is perhaps more about reconfiguration than niche development. However, this is increasingly intertwined with the niche expansion and mainstreaming of wind power as well as continued support for established nuclear power. Both electrification and the larger share of wind power in electricity production have changed the key logic of the Finnish energy regime to become more responsive to energy availability at any given time (with development in demand and consumption response) and have reduced the role of stockpiling. The objectives behind these developments not only relate to decarbonization but also to visions of Finland as a net-electricity exporter and as attractive to new industries.

### 6.5.1 Wind Power and Its Conflict with the Defence System

In the beginning of the period of interest for this study, around 2006, the potential of wind power to improve energy security was hardly mentioned in Finland (Varho, 2007). In later discussion and policy documents, the negative effect of wind turbines on Defence Forces' prewarning systems, namely air surveillance radars, emerged as a significant issue. The operation of military air surveillance radars requires that the radio waves they send meet with no obstacles between the radar and the target, the reach of the radar being several hundreds of kilometers and 0–20 kilometers in height; wind power turbines reduce the radar range and the blades create reflections on the radars (Joensuu et al., 2021). As a consequence, the restrictions for wind power construction in eastern Finland has meant that wind power is concentrated mostly in one part of the country and the energy system is less able to exploit varying weather conditions in different parts of Finland.

During 2008–2009, certain wind power developments were halted due to the opposition of Defence Forces and the unclear effects on the radar systems (Joensuu et al., 2021). In many eastern Finland regions, close to the Russian border, wind power construction has not been allowed. The Defence Forces have employed case-by-case consideration of wind power plans. Due to highly confidential factors related to defence, those planning wind power projects and those responsible for land-use planning do not have the information on how to design these projects in order to make them more acceptable; from the perspective of the Defence Forces, the project proposals are either accepted or rejected and little to no guidance has been available for wind power developers (Joensuu et al., 2021). However, continued negotiations have been possible in some cases (MEE, 2014).

To better enable wind power construction, a methodology was created during 2010–2011 by the VTT Technical Research Centre of Finland on assessing the wind farms' effects on radars; this development was funded by a group of twenty wind power companies and the public administration and was coordinated by the Energy Industries Federation. As a result, by spring 2021, the Defence Forces had approved over 600 wind power developments composed of 11,000 wind turbines. Of these, 128 were in regions located in the eastern side of Finland but only 28 in the southeast border regions. One of the key problems has been that obtaining a statement from the Defence Forces has been a lengthy process and is not defined in law (Joensuu et al., 2021).

In 2012, the MEE established a working group with the task of advancing the shift to wind power, removing barriers to wind power construction, and coordinating the objectives of different ministries. An arrangement was also made by which obstacles to the operation of radars caused by wind power could be removed by developing the Defence Forces' radar systems. Effectively, in 2013, a new law on wind power compensation areas removed barriers for wind power construction in the west of Finland for offshore wind on the Gulf of Bothnia, covering 2,400 square kilometers. This meant that construction was possible in previously prohibited areas. However, the Defence Forces required that compensation, effectively additional sensors, could not be funded from the defence budget and the wind power producers in these areas need to provide so-called radar compensation payments to cover the additional costs of new radar technology. In turn, a similar compensation area in the Kymenlaakso region closer to Russian border was found to be so expensive to implement that it was regarded economically unviable (MEE, 2014).

Generally, it has been perceived that the defence administration's perspective on wind power is understandable and acceptable, the Defence Forces are not against wind power as such, and the administration has been engaged in processes that seek to enable wind power construction. The Defence Forces have, since 2009, actively collaborated with other administrative sectors and wind power developers (MEE, 2014). Common solutions have been sought, as demonstrated by the compensation scheme, yet viable technological solutions in the eastern areas have been hard to find.

Following the 2022 war in Ukraine and the energy crisis, Finland increased its support for the expansion of wind power to replace the shortage of electricity previously imported from Russia. This meant new policies speeding up wind power-permitting processes and appointing an official expert to assess potential technical and operational solutions that would allow the expansion of wind power and still enable effective air surveillance in eastern Finland. Some perceived that these new policy lines on wind power would not have been taken forward a few years ago. Nevertheless, it appears that no technical solutions exist to expand wind power substantially further east, and there needs to be a 40–80-kilometer distance between a wind park and a radar (Pöntinen, 2023). Therefore, the conflict between expanding wind power in eastern Finland and maintaining national security prevails.

In this case, state (defence) as a referent objective and negative security in the current world situation have, understandably, taken preference over improved energy security or regional viability (as elements of positive security). Although the compensation model has benefited certain regions, it has placed eastern regions in less attractive positions with regard to new industrial investment and sustained livelihoods. Learning processes and networking have taken place both administratively and technologically, but comprehensive solutions to the policy conflict have not been found despite expectations that wind power will be massively expanded in the coming years. Further, the expansion of wind power, while improving energy security, will also create new dependencies on components and critical materials – linked to geopolitics and trade with China. The events of 2022 led to greater interest in energy transitions policy for wind power, but not to a degree that would have substantially securitized the decision-making.

### *6.5.2 Peat Energy and Its Promotion as Traditional Energy Security*

In Finland, peat and bioenergy are the only domestic energy fuels and, hence, are tied to the decline of the fossil regime. The energy use of peat began as a result of the first oil crisis of the 1970s, with peat providing a domestic energy source to replace some of the imported fossil fuels – especially oil – for heat and power production via combustion. As a new energy source at the time, it was well suited for storage, contributing to security of supply. When climate change concerns began to increase, replacing peat with bioenergy in combustion plants came into focus. As noted in the introduction to this chapter, the share of peat in total energy consumption is relatively small but makes a rather substantial contribution to Finland's total greenhouse gas emissions. Large wetland areas have enabled the use of peat for energy production, supported by favorable long-term policies, such as subsidies in different forms, making peat more competitive in relation to other energy sources (IEA, 2018). This has slowed down Finland's zero-carbon energy transition. Following the EU Emissions Trading Scheme, the Finnish government created a peat promotion scheme for 2007–2010 that allowed an additional premium tariff (paid as a subsidy on top of the market price) for electricity produced with peat and extended a tax exemption for peat used in heating (IEA, 2018). In 2017, the government's "Climate and Energy Strategy" noted that taxation will still be used to ensure that peat remains more cost-effective than imported fossil fuels (MEE, 2017).

Peat is regarded important for the local economies and employment in certain regions of Finland, hence contributing to internal socioeconomic security. It has also played a role in energy security via stockpiles equaling six months' use. VAPO, a government company established in the 1940s, initially as "the state's fuel office," has been a significant producer of peat from lands owned by the government, with private sector ownership increasing in the 2000s. It has been an active campaigner for peat energy. In a campaign launched in 2010, following the Russia–Ukraine gas delivery disruptions, it lobbied in favor of peat by using, in particular, the energy security argument, taking advantage of the repoliticization of energy security (Lempinen, 2019). Peat has essentially been linked to a storyline that regards "the energy transition and security as incompatible" (Höysniemi, 2022).

Interestingly, peat as an energy security question was mostly brought up by the politicians interviewed and not by other actors. It illustrates the politicization of the peat question over other energy questions and how the security aspect of peat is linked to political tactics. What seems clear is that security of supply has been used as a lobbying tactic as well as a traditional energy security argument in favor of peat, not by the ministries or Defence Forces, but by NESA, as pointed out by one interviewee. The importance of peat for security of supply was, however, argued by another interviewee to be reducing in the transition from combustion to other forms of electricity and heat production. Yet a third interviewee noted that "peat is important, no matter what anyone says."

The phaseout of peat in energy production has largely been market-based, resulting from the EU emissions trading system. This development, however, experienced a setback in 2022 in response to the security and energy crisis. The crisis led to halted wood imports from Russia, while there was also a concern that domestic wood was being used for energy production instead of higher-grade uses in the forest industry. Therefore, tax-free use of peat was allowed up to 10,000 MWh per year during 2022–2026. Furthermore, the government's "Climate and Energy Strategy" from 2022 pointed out, for instance, that the reduction of peat and fossil fuels directs attention to the heating sector's security of supply. It also outlined a new policy, an emergency stockpile of peat used in energy production. Peat production areas are also a focus of the EU just transition efforts and are connected to positive security. To improve socioeconomic security in regions dependent on peat production as a source of employment, funding and advice are made available to develop new businesses, find new employment, and compensate for the depreciation of peat production machinery (MEE, 2022).

Peat as an example of the energy–security interplay places the image of "traditional energy security" as a reference object alongside, perhaps, regional viability – if the latter can be regarded as a referent object for security. Peat, therefore, highlights the negative security aspect of the energy transition, while emphasizing regional livelihoods as positive security. While the latter is important for security, the peat debate includes no long-term approach toward achieving climate and geopolitical security in a rapidly changing world.

### 6.5.3 Nord Stream 2 Gas Pipeline

The ways in which the Finnish government dealt with the Nord Stream 2 gas pipeline application – as it went through Finland's territorial waters – came up frequently in the expert interviews related to energy and security. This case did not involve Finland's energy consumption as such, but it was linked to maintaining fossil fuels in Central Europe and Finland's relations with Russia. Hence, it is covered here briefly.

Nord Stream refers to a network of offshore gas pipelines constructed to supply Russian natural gas to Germany. The company operating the pipelines is majority-owned by the Russian government-owned company Gazprom (with 51 percent), with the remaining shares owned by German, Dutch, and French companies. The two pipelines of Nord Stream 1 were completed in 2011 and of Nord Stream 2 in 2021.

Nord Stream 2 is an example of officially not connecting security to a fossil fuel project important to Europe, when perhaps this should have been done. Several experts noted that in essence the Nord Stream pipeline project was depoliticized on purpose in Finland. An interviewee remarked that there was an avoidance of discussion: The use of the term "security policy" in particular in that context was avoided and a desire existed to view the gas pipeline as a neutral project despite its potential security ramifications. Certainly, Russia regarded it as security policy, another interviewee stated. Yet, essentially, a "political-level" decision was made to not address this issue as a security question and Finland did not take part in international geopolitical discussions.

Officially, the project was addressed as something with potential environmental impacts on the seabed that were assessed in the consultation phase. This perspective was tied to the international agreements of which Finland is a part and the stipulations they have in place. These did not include geopolitical aspects.

Several explanations were offered for the lack of discussion around security. Several experts took the view that the project was important for Europe–Russia relations at the time – and essentially opposition was framed in terms of the economic interests of the US to provide Europe with LNG. It did not affect Finnish energy policy as such and could curtail the energy company Fortum's ventures in Russia and Germany. One interviewee said that Finland has "a tendency to deal with these kinds of issues calmly" and two referred to the lack of any previous problems.

Given the lack of security discussion, it is hard to define a reference object – perhaps the European energy system or wider geopolitical security. As the debate was depoliticized and desecuritized, any positive and negative security implications remained implicit. Security-of-energy supply for Europe could be seen as an element of positive security. In turn, Europe's increased dependence on Russia, climate change effects, and possible infrastructure risk (realized in 2022) were elements of negative security.

## 6.6 Concluding Remarks

Before 2022, there was an explicit and perhaps politically intentional disconnection between energy policy and geopolitical security. This has also been noted by others (Jääskeläinen et al., 2018). The strong economic interests of Finland for cheap energy imports for industry and of the Finnish energy company Fortum – together with the history of Finlandization, that is, maintaining friendly terms with the Russian government – prevented a genuine assessment and discussion of the geopolitical risks related to energy trade with Russia. Interestingly, Finland was able to maintain a strong defence policy at the same time, which has also kept many areas of eastern Finland, close to the Russian border, inaccessible for wind power expansion. All in all, Russia has been a major landscape influence on the Finnish energy transition – somewhat slowing it down. This effect has occurred via the aim to achieve inexpensive energy for the export industry – where Russian imports have often been low priced – while, simultaneously, decarbonization of the electricity sector has been rather strongly pursued, creating policy incoherence.

Some energy system aspects, such as peat and nuclear power, have been subject to relatively intensive political debates. Yet, at the same time, a goal to depoliticize

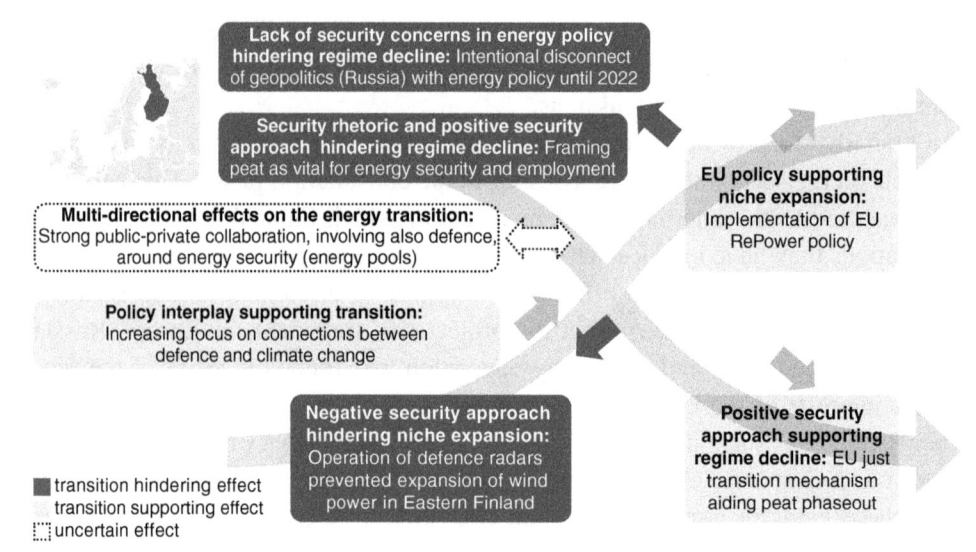

Figure 6.5 Key energy security aspects and their transition impacts in Finland. Source: Kivimaa, Finnish Environment Institute, 2023.

those parts of the energy system that were linked to Russian trade has existed. Examples of this include the general reluctance to discuss geopolitical security – while economic perspectives on security of supply existed – the depoliticization of the Nord Stream 2 gas pipeline debate, and the aim to depoliticize nuclear power. High trust by citizens in expert-based policymaking has supported the depoliticization of some energy questions. Figure 6.5 summarizes the key energy security issues in Finland and their effects on the energy transition.

The events of 2022 led to a substantial shift in perceptions of Russia with regard to the energy system context, the politicization of energy policy, and some exceptional energy policy changes. Time will show whether this will also be influential in changing the energy governance system and how. How will energy and geopolitical security be addressed by the next governments and will that result in a less fragmented energy-governance system? How will the negative and positive security implications of energy transitions be anticipated and reacted to? The experimental governance features emphasized by sustainability transitions studies and studied and tested in other Finnish contexts have been rare in the rather "old world" energy administration of Finland but could be of use here (Kivimaa and Rogge, 2022). Certainly, issues linked to electrification and the expansion of wind power necessitate much governance experimentation and innovation to deal with the defence radar issue as well as the sustainable extraction and supply of critical materials.

# 7

# Norway

*Contradiction of Oil for Export and Fully Renewable Electricity Supply*

The Northern European country Norway has a population of 5.5 million and is among the wealthiest countries in the world. Norway was part of Denmark until the nineteenth century and gained independence from Sweden in 1905. The country is one of the founding members of the North Atlantic Treaty Organization (NATO) and, while it chose not to join the EU, it is required to follow some EU legislation and has access to the EU internal market via the European Economic Area (EEA) Agreement, signed in 1992. Norway is located on the western side of Sweden and has limited borders with Russia and Finland to the north. On the northern, southern, and western sides, Norway is surrounded by the Barents Sea, the Norwegian Sea, and the North Sea. Its coastline is over 2,000 kilometers. The continental shelf governed by Norway is four times the size of the mainland and one third of the whole European continental shelf. Therefore, Norway has substantial access to offshore oil and gas reserves located on its continental shelf. It has also deliberated opening an enormous area for deep sea mining of critical minerals (Alberts, 2023).

Norway is a unique country case as it holds such a large magnitude of different kinds of energy reserves: oil, natural gas, and hydropower. These reserves have made Norway fully energy independent in terms of domestic energy consumption, apart from some dry years when it imported electricity to compensate for the lack of hydropower (Figure 7.1). Norway exports about nine tenths of its energy production (Figure 7.2). The country generated 3 percent of global gas (being the seventh-largest producer) and 2.3 percent of global oil production in 2020; the International Energy Agency (IEA) has described Norway as having a politically and economically "stabilizing role in the world's oil and gas supply" (IEA, 2022, p. 9), which is otherwise concentrated in the Organization of the Petroleum Exporting Countries (OPEC) and the US. The income generated from the export of oil and gas has created substantial societal gains for Norway, but has also meant the country is economically dependent on this income. Despite increasing climate

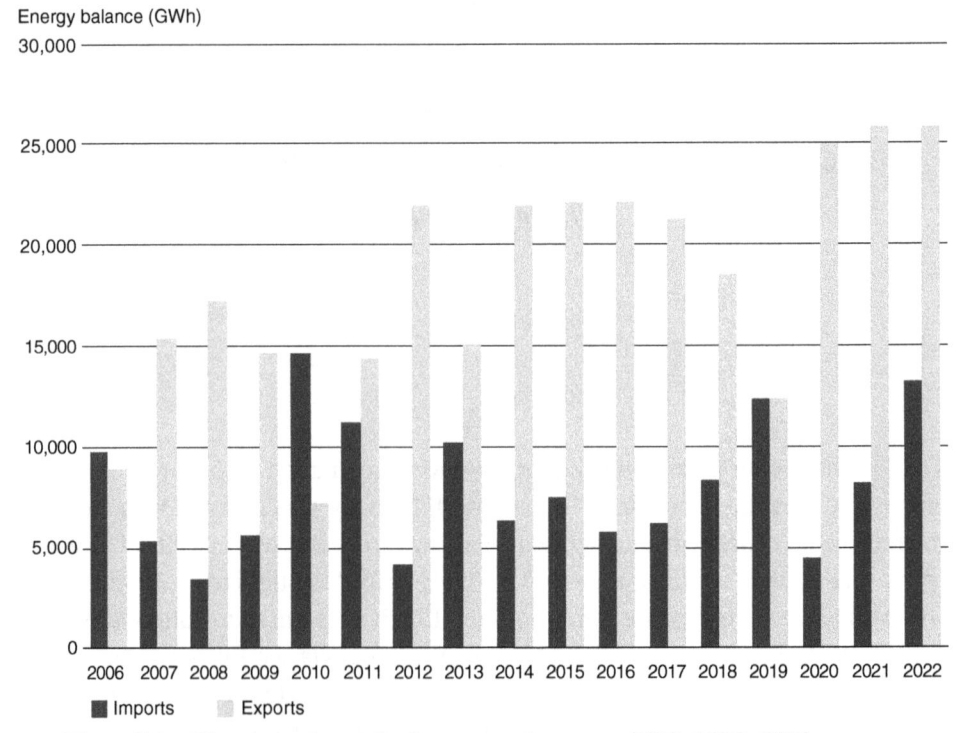

Figure 7.1  Norway's electricity imports and exports, 2006–2022, GWh.
Source: Statistics Norway (2023a).

concerns and risks associated with oil and gas production the governance structures of Norwegian hydrocarbon production remained unchanged during 2013–2018 (Bang and Lahn, 2020).

Besides hydrocarbons, Norway has ample waterfalls that have, over several decades, led to the development of a sizeable hydropower sector that covers nine tenths of Norway's total electricity production. Therefore, as Norwegian society is highly electrified and decarbonized, other renewable energy sources have been rather poorly supported politically. Instead, the oil and gas sector were supported by, for example, new concession rounds for exploration during 2015–2017 (Mäkitie et al., 2019). Unlike the other case countries in this book, for Norway, the renewables niche expansion via wind power has often been perceived to mean rising levels of exports and power exchange to other countries without improved security of supply (Hansen and Moe, 2022). Wind power can, however, play an important role for local security of supply in areas where connections to the transmission network are weak (Skjølsvold et al., 2020). Whereas wind power was perceived positively in the early 2010s (Karlstrøm and Ryghaug, 2014), subsequently it faced increasing citizen resistance for various reasons, including the dispossession of Sámi areas

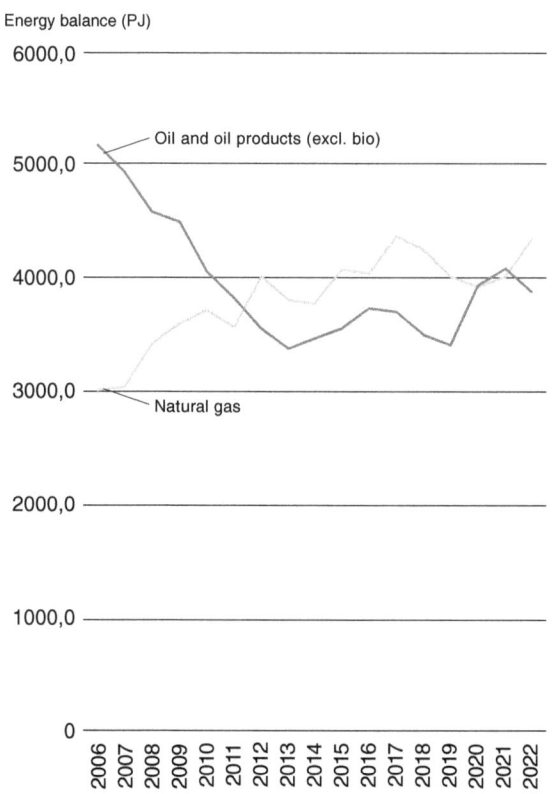

Figure 7.2 Norway's export of oil and gas, 2006–2022, in millions of tons oil equivalent.
Source: Statistics Norway (2023a).

(Normann, 2021) and the rise of resource nationalist arguments (Hansen and Moe, 2022). It is only since the 2020s that increased policy support has been given to wind power to accelerate its expansion (see details in Section 7.5).

The abundance of fossil energy has led to a kind of dissonance between Norway's export-oriented fossil fuels policy and its domestic zero-carbon energy policy, which are dealt with administratively in different domains and ministries, perhaps to reflect the different values involved. The fossil fuel sector represents economic security for Norwegians, while zero-carbon energy reflects the country's environmental values. Therefore, to limit the emissions of the fossil fuel sector, Norway has developed its hydrocarbon production to be the world's least carbon-intensive and has invested heavily in the development of carbon capture and storage (CCS). Nevertheless, this dissonance has meant that energy security concerns played a substantially smaller role in Norway prior to 2022 than in the other case countries.

The war instigated by Russia in Ukraine and the resulting energy crisis in Europe have affected Norway differently than other European countries, due to the country's large energy reserves. The events nevertheless caused changes in Norwegian energy policy too. The impact was twofold. First, Norwegians experienced much higher electricity prices than before due to interconnections with other European countries (Germany, Finland, Sweden, the UK, the Netherlands), although the government compensated consumers for this rise to a large extent. Second, the Norwegian government was able to benefit from record high exports of oil and gas, replacing the quantities previously supplied by Russia to Europe and receiving thanks from the EU President Ursula von der Leyen, for being a valuable source of help in the European energy crisis (von der Leyen, 2023). This surplus in income accrued to the whole of Norwegian society because of large state ownership in the fossil fuel sector and a specific fund that saves and reinvests these profits for the benefit of the country and its future generations. Effectively, the way in which the Norwegian fossil fuel sector is governed represents thinking associated with positive security. The approach in which social benefits come from the fossil fuel funds has enabled, for a long time, the creation of high living standards and an economically and socially stable society. It has, however, also created an economic problem related to the potential future phaseout of fossil fuel production, on which no decisions have been made.

The few security concerns regarding the energy system in Norway, prior to 2022, focused mainly on the safety of hydropower installations (see Section 7.5). An area of focus in Norway concerning oil and gas has been control over the Svalbard area where many hydrocarbon resources have been of interest to both Norway and Russia. Additionally, the northern Finnmark region has a border with Russia and, hence, is important for the country's territorial defence.

As with previous country chapters, this chapter describes the key context, that is, the energy and security regimes in Norway. It then continues with the analytical sections, drawing on Chapter 4, namely the perceptions of Russia as a landscape pressure for the energy sector; policy coherence and interplay between energy and security regimes including the level of securitization; and, finally, positive and negative security related to potential niche development and regime (de)stabilization. The analyzed material covers Norway's energy- and security-related government strategies published since 2006 and two rounds of interviews with energy and security experts, first between November 2020 and March 2021 and then again between November 2022 and March 2023. The chapter draws from these materials and related literature and selected policy reports.

## 7.1 Energy Regime

As noted, Norway has a high share of renewable energy in its total energy consumption and its electricity sector is practically fully decarbonized. In 2022, 88 percent

## 7.1 Energy Regime

of Norway's electricity consumption was produced with hydropower (reduced circa 10 percent from previous year due to dryer conditions) and 10 percent from wind power (an approximately 25 percent increase from 2021) (Statistics Norway, 2023b). Contrary to an earlier slow development of wind power, particularly due to bottlenecks in the concession process (Blindheim, 2013), wind power production doubled between 2019 and 2020, increasing threefold by 2022. Despite Norway already being a leading country in the electrification of its society, there are new demands for further electrification, for instance, from the industry and petroleum sectors, data centers, green hydrogen production, and transport (IEA, 2022). This is likely to affect electricity prices and creates a need to expand wind power capacity.

If one looks at total energy production instead of domestic consumption, a different picture emerges. Oil and gas, with an even share, constitute nine tenths of energy production and the remainder is largely hydropower. There are no explicit and immediate plans to phase out oil or gas production. This is due to the economic security these provide for Norway, but also to broader European energy security and arguments that Norwegian oil and gas are produced in a more environmentally friendly way than elsewhere. Still, uncertainty exists regarding future demand for fossil fuels from other countries, even though the events of 2022 stabilized Norway's role as an oil and gas provider for Europe in the short term and resulted in plans to expand current production fields. In addition, future energy regime development in Norway is described in terms of visions for both blue hydrogen (fossil fuels) and green hydrogen (from renewable electricity generation).

The largest fossil fuel producer in Norway is Equinor. In 2022, it produced over two million barrels of oil equivalent per day and circa 1,650 gigawatt hours of renewable energy (Equinor, 2023). When energy transition began gaining ground, in 2018, Equinor changed its name from previous Statoil and began new renewable energy projects. The state owns 67 percent of Equinor, and it is operated by the Ministry of Trade, Industry, and Fisheries. Other oil and gas producers are less than half of the size of Equinor. The largest of these include Esso Norge (part of Exxon Mobil) and Total E&P Norge.

The largest electricity-sector companies include Norsk Hydro, Statkraft AS, and Statkraft Energi AS. The Norwegian state owns 34 percent of Norsk Hydro; with the rest of company owned by private investors. Statkraft AS has multiple subsidiaries, all fully owned by the Norwegian state. It was described by an interviewee as Europe's largest renewable energy company. Also, municipal ownership of hydropower is significant. Contrary to hydropower, most wind power is foreign-owned. Statnett, the transmission network operator, was established in 1990 when the electricity market was liberalized. In contrast to companies in the other countries described in this book, Statnett is fully owned by the Norwegian state, more specifically by the Ministry of Petroleum and Energy.

Norway's large public-sector ownership of energy companies is nowadays quite a unique feature and enables tight collaboration between key energy-sector companies and the government. In the administrative side, there are several ministries that can be regarded as key actors pertaining to energy and security. The Ministry of Petroleum and Energy deals with the exporting oil and gas sector, while the Ministry of Trade, Industry, and Fisheries is the majority shareholder in Equinor and Statkraft SF. The former has overall responsibility for energy and water resources. The Ministry of Climate and Environment is relevant from the perspective of the climate and the environmental effects of the energy sector. A few experts interviewed described the Ministry of Climate and Environment as weak compared to the energy ministry, which was seen as very strong:

The Ministry of Petroleum and Energy is a very strong and entrenched ministry with a long tradition. It's come of age alongside Equinor and the industry, and it's over the years clearly been seen as strongly promoting the exploration of oil and gas on the Norwegian continental shelf and then increasingly abroad. (Business actor, 2021)

The Norwegian Water Resources and Energy Directorate (NVE) operates under the Ministry of Petroleum and Energy. It is responsible for the management of energy resources and is the regulatory authority regarding the electricity sector. Interestingly, some experts described the NVE as almost as a ministry itself and pointed out that oil and gas exports are in practice separated from electricity production. Such an arrangement can be seen as, on the one hand, advancing the zero-carbon transition of the electricity sector and, on the other hand, as maintaining the oil and gas export sector:

That is a kind of a strange feature of Norway ... so you have one ministry for petroleum and energy, but in practice it's two ministries, or it has been so for a long time where we have separate processes for electricity and for petroleum. (Researcher, 2021)

ENOVA SF is a state-owned enterprise that is tasked with the advancement of more environmentally sound energy-sector development and supervised by the Ministry of Climate and the Environment. It does this by managing the Climate and Energy Fund, which aims to realize projects advancing Norway's climate commitments.

Norway is not an EU member state but is still bound by many EU policies: It participates in the internal energy market as part of the EEA Agreement as well as in EU climate legislation for the period 2021–2030 (IEA, 2022). The EU is also Norway's largest energy export market. An interviewee, however, remarked that Norway lacks similar holistic climate and energy policy preparation processes to many EU member states. Indeed, compared to many EU countries, renewable energy beyond hydropower has received little policy support in Norway. Renewable energy support was first highly politicized, but from 2010 its status as a salient political issue reduced: "Norway was producing more electricity than it needed domestically; electricity prices dropped, and the electricity utilities changed their position towards the certificate scheme" (Boasson, 2021, p. 205).

Norway aims to reduce its greenhouse gas emissions by at least 50 percent by 2030 and by 90–95 percent (from 1990 levels) by 2050. It also has a goal for carbon neutrality by 2030 via domestic reductions, the EU emission trading system, and international cooperation on emission reductions (IEA, 2022). The IEA assesses that Norway has already gone down the easiest routes, such as electrification of many parts of the society, so further greenhouse gas reductions will be more difficult to achieve (IEA, 2022). The Climate Change Act was enforced in 2017 to make the emission targets legally binding. In addition, the 2050 Climate Committee was established to conduct an overall review of Norway's choices with regard to achieving its 2050 climate target.

In 2021, the Norwegian government published a White Paper "Putting Energy to Work," which aimed to continue fossil fuel exploration indefinitely. It, however, also described a strategy to repurpose the skills and assets of the Norwegian oil and gas industry toward the development of new industries and technologies, largely offshore wind and green hydrogen but also CCS and blue hydrogen. Therefore, it seems that Norway is on board with the renewables niche expansion but, due to high economic security, it cannot endorse the decline of the fossil fuel export regime. Following the 2022 events, the Norwegian government issued a supplementary White Paper. The updated White Paper stated more strongly than before that policy for the further development, not discontinuation, of the petroleum industry needs to be put in place.

Indeed, since 2022, according to the interviewed experts, Norwegian energy policy has changed substantially. One significant difference is the strengthening of Norway's role as a stable and credible energy provider to Europe. In 2022, Norway supplied 8 percent of Europe's oil consumption and it continues to play an important role producing oil and gas for Europe. Existing oil and gas production areas were opened for further exploration. At the same time, wind energy is of interest too, and new areas were opened for offshore wind development. However, an interviewee expressed a concern that the oil and gas sector, as a higher salary-paying sector, will attract the majority of the skilled workforce, with less workers available for the offshore wind sector. This has proved to be the case before too (Mäkitie et al., 2019).

The new plan for wind power expansion could be seen as a break from previous political practice (Kuzemko et al., 2022). One of the latest developments includes building the world's largest floating offshore wind farm (Hywind Tampen), based on Equinor's floating wind technology, with a total installed capacity of 88 megawatts (IEA, 2022, p. 12). Equinor has also invested in wind power production outside Norway. One of the bottlenecks in Norway is the status of transmission and distribution networks. An interviewee argued that the grid is ruled by institutions that were set up following energy market deregulation in the 1990s and geared toward curbing grid investments. Nevertheless, Norway's industrial policy is pushing for a new direction that focuses on access to renewable energy, battery production, hydrogen production, and new energy infrastructure for shipping:

There has been a strong drive for this green transition. On the one hand, we are producing as much oil and gas as we can, and on the other side, we're trying to make us be looked at as the greenest country ever. It's a schizophrenic position somehow. But the war in Ukraine has increased the importance of oil and gas at least for a period of time. The efforts being done in order to position Norway as a key contributor of technology, when it comes to onshore wind, onshore wind, is very much increasing, and it's a very high priority for the Norwegian government. (Civil servant, 2023)

Norway has the lowest-emitting oil production facilities globally. It heavily invests in CCS to reduce the greenhouse gas emissions of fossil fuel production, driven by the carbon tax introduced in 1991, with Equinor implementing several CCS projects (IEA, 2022).

In February 2022, the Norwegian government appointed an Energy Commission to assess future challenges in Norwegian energy policy up to 2030. In the resulting report in spring 2023, the commission called for a change of pace and the establishment of new green industries. If power production is not increased, the commission warned, there is risk of an energy deficit – potentially as early as 2027. Figure 7.3 summarizes the key aspects of Norway's energy policy.

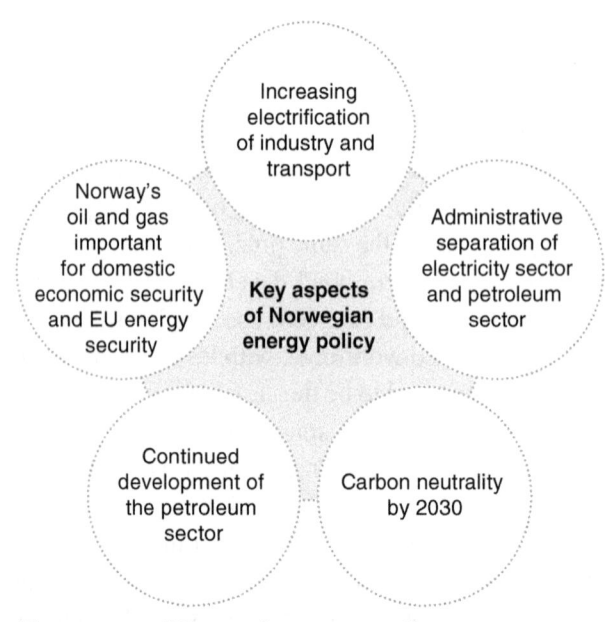

Figure 7.3  Key aspects of Norwegian energy policy.

## 7.2 Security Regime

Norway applies a total defence concept in its defence policy. Total defence has been described as integrated military and civil preparations supported by institutionalized cooperation between ministries, civic organizations, the private sector, and the general public (Wither, 2020). The Directorate for Civil Preparedness heads up the Total Defence Working Group and coordinates exercises. A key aspect of Norway's defence policy has been its membership of NATO. Norway was one of the founding members in 1949. Also, akin to Finland and Estonia, Norway now considers Russia to be the main threat to its national security. Before 2022, Norway (as well as Finland) collaborated with Russia on many economic, environmental, and border security questions (Wither, 2020).

Norway also collaborates with the EU on defence and security. This collaboration has been described as ad hoc and based on informal arrangements (Hillion, 2019). In terms of formalized collaboration, Norway has relatively little influence on EU defence and security policy, although it indirectly plays a role as a NATO member country and as an Arctic state. In relation to security and foreign policy, Norway emphasizes the importance of international collaboration, NATO membership, and multilateral systems guided by the UN (Norwegian Ministry of Foreign Affairs, 2019).

The relevant ministries include the Ministry of Defence, the Ministry of Justice and Public Security, and the Ministry of Foreign Affairs. An interviewee reported some rivalry between the ministries of defence and of foreign affairs. This was prior to 2022 and pertained especially to their orientation toward Russia – the Ministry of Foreign Affairs being more open to collaboration. In the Norwegian parliament, Storting, the Committee on Defence and Foreign Affairs deals with security issues.

Norwegian defence spending gradually reduced by half from the 1960s to 2000. It was at its lowest in 2008, 2012, and 2013, representing 1.4 percent of the GDP. In 2021, 1.8 percent of GDP was used for defence, which corresponds to the lowest percentage of this book's case countries, but, given Norway's substantial wealth, in monetary terms this was not the lowest amount.

The national security legislation was revised in 2019. The national defence plan is renewed every four years, and it includes a section on energy security. There is some overlap via organizational roles: Besides its energy policy roles, the NVE (see Section 7.2) develops guidelines for cybersecurity in the energy sector and conducts inspections. A previous analysis of Norwegian security and defence policy documents showed that policies have highlighted the Norwegian responsibility to govern the sensitive and important Arctic areas, while globalization was expected to increase cybercrime and terrorist attacks (Sivonen and Kivimaa, 2023). Figure 7.4 shows the key aspects of this sector.

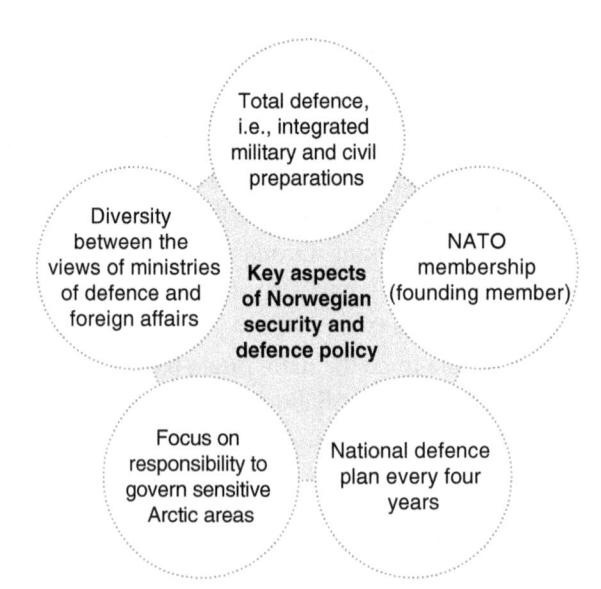

Figure 7.4  Key aspects of Norway's security and defence policy.

## 7.3 Perceptions of Russia as a Landscape Pressure at the Intersection of Energy and Security

Norwegian policy documents relate principally to the importance of international collaboration and trade. For example, developments such as China's Belt and Road Initiative, the proliferation of nuclear materials, world population growth, increasing resource demand, and climate change are seen as important landscape developments affecting the energy sector (Norwegian Ministry of Foreign Affairs, 2019).

In terms of Russia as a particular landscape pressure, Norway has not been dependent on its oil and gas, unlike Estonia and Finland. Otherwise, the perceptions and expectations in Norway prior to 2022 presented a similar dual approach to Finland, both in policy documents and interviews. Before 2022, about a half of the interviewed experts either did not really consider Russia with regard to the energy–security nexus or perceived relations with Russia to be neutral or good. Of the remainder, one quarter presented mixed perceptions of Russia as a landscape pressure. For instance, the geopolitical threat and increasing military activity of Russia was mentioned but, at the same time, no threat toward energy installations or collaboration in environmental issues was perceived.

There are some nuances about how to balance this so we have some voices that are more concerned about provoking Russia and others are saying that no we should be firmer. That's a debate we've had since 2014 how to balance this. We don't want to escalate things up North, but we still want to be firm. (Researcher, 2020)

Only three interviewees perceived potential Russian developments as a risk. Two mentioned Russia's increased assertiveness. In addition, risks were perceived in relation to the potential environmental implications of oil and gas transport and nuclear power. Further, three interviewees reported the 2014 invasion of Crimea as a major shift in Norwegians' perceptions of Russia.

We have experienced a more assertive Russia. A Russia that has become quite harsh when it comes to their characteristics on what's happening. And then it's, it's also a difference on whether Russia speaks to Norway as a NATO member or whether it speaks to Norway as a neighbor. (Civil servant, 2020)

Russia has become more assertive. They have increased their military reach, and they are very critical of whatever defence measures Norway decides to take, and our cooperation with the Americans and all the allies. So, after Crimea and the sanctions in 2014, Russia has increasingly become untrusted. (Business actor, 2021)

Although Norway's connections with Russia have been quite different from Finland's, that is, Russia is not an energy importer to Norway but rather shares interests in the Arctic Sea, nevertheless, perceptions of Russia were similar and there was an emphasis on collaborative relations. As in other case countries, 2022 gave rise to a major landscape shock, resulting in more uniform views about Russia in the energy–security nexus.

Norway is supporting all EU sanctions. Norway is supporting any statements from NATO and the European Union and the United Nations. So, we are rock solid on the Western perception of Russia. (Civil servant, 2023)

Well, I think Russia is now considered a pariah. Even a rogue state. Nobody in the West, with the possible exception of Viktor Orbán, will ever trust Russia again in energy matters, as long as Vladimir Putin remains in the Kremlin. (Business actor, 2023)

I think our perception at the moment is that they're not somebody you can plan with at all, for the European market. (Business actor, 2023)

Next, the chapter moves onto discussing policy coherence and interplay at the energy–security nexus.

## 7.4 Policy Coherence and Interplay

As Norway is a special case of a country fully independent in energy and a major energy exporter, the ways in which security issues unfold in its energy transition differ substantially from that of Finland and Estonia. Norway is not dependent on the import of any energy source and, hence, energy security in Norway does not relate to security of supply but rather broader energy security: the operation of hydropower plants and the potential disruptions they may face, either from hybrid attacks or weather events. As noted in previous literature, the mainstream energy policy discourse "does not see renewables as connected to energy security, independence, or sovereignty" (Hansen and Moe, 2022, p. 4). In practice, however, this is not so straightforward. There are limitations in transmission capacity between different regions in Norway, which mean that availability and prices of electricity can alter substantially between regions. These issues with the grid infrastructure were not, however, presented as a security issue in Norwegian interviews.

During the expert interviews taking place in 2020 and 2021, security was seldom mentioned as connected to energy in general or energy transitions more specifically. This was due mostly to the energy independence of Norway; the security of the energy infrastructure alone was identified to be potentially at risk (of terrorist attack), yet most regarded this as unlikely. Some references to energy security were made, as potential electricity shortages in dry years would affect hydropower. In addition, cybersecurity was identified as an issue for the energy sector.

Likewise, in the first round of expert interviews, there was little consideration of how energy and security issues cohere with each other in public policymaking. There appeared to be "no committees, working groups or agencies advancing coherence between zero-carbon energy and security" (Kivimaa, 2022, p. 8). However, there were features that connect to the energy–security interface. For instance, emergency preparedness procedures and guidelines for cybersecurity by the NVE have been created in this interface. The NVE also heads up the Power Supply Preparedness Organization. Similarly to Finland, this organization, responsible for national emergency strategy, comprises representatives from different ministries, including those in charge of energy and of foreign affairs (IEA, 2022). The Petroleum Safety Authority has responsibilities regarding the security of petroleum production. Nonetheless, the overall interpretation of expert interviews was that, prior to 2022, energy and security policies were not coordinated between the four administrative organizations involved: three ministries and the NVE.

My main understanding is that this combination does not almost exist in Norway. (Business actor, 2021)

The oil and energy department has been quite adamant to demonstrate an independence from foreign affairs and security policy. For certain reasons, Norway has its foreign relations, it has its defence and security set up, but apart from the obvious things of securing oil installations and so on, we have tried to keep energy policy separate, not to politicize. (Politician, 2021)

In principle, fossil energy has been depoliticized in connection to the importance of this sector for the Norwegian economy. This has also impacted how its coordination – or lack of it – was pursued in foreign and security policy (Kivimaa, 2022). On the domestic side, the security of hydropower installations has attracted attention, especially since a cyberattack on a Norwegian hydropower company in 2019 (see Section 7.5.2). In 2023, Storm Hans showed how extreme weather events resulting from climate change can also be a threat to energy infrastructure. Increase in the water level of the Glåma River destroyed a dam and inundated a hydroelectric power plant (Patel, 2023).

There have also been differing views, such as the NVE having close connections to military forces and trials around total defence, and the energy ministry having a subsection conducting security analyses. Moreover, informal interactions via a small elite in a small country were mentioned as justification for the lack of more formalized processes – akin to Estonia. In 2018, the Council for Fuel Preparedness was formally established to increase cross-governmental cooperation on fuel security (IEA, 2022). Nevertheless, synergies between energy transition policy and security policy were not recognized, whereas the fossil fuel sector was identified as a positive factor for Norway's economic and geopolitical security (Kivimaa, 2022).

Since 2022, more coordination has occurred, although an energy business actor noted that actors outside ministries have little awareness of whether and what kind of interaction takes place:

Equinor as the country's major corporation has in its 50th year of operation finally in part come under the jurisdiction of the National Security Act ... I think many of my colleagues will argue that this was long overdue. (Business actor, 2023)

Among the interviewees for this book, energy business actors appeared to have a broader awareness of security connections than civil servants. Since 2022, the security of electricity supply – or at least the affordable price of it – became a key part of political discussions. This was the result of new electricity export interconnections from Norway to other countries in 2021 and an overall reduction in energy supply from Russia to Europe in 2022. The updated energy policy from spring 2022 was much more explicit about security than previous strategies. It was entitled "Energy Policy for Employment, Transition and Security in Times of Uncertainty" and emphasized Norway's continued and stable oil and gas production. Given previous obstacles to develop wind power further, the measures to speed up wind power development can, to a certain degree, be seen as exceptional measures (cf. Heinrich and Szulecki, 2018). Future developments will show whether the coherence between energy and security policies will improve.

## 7.5 Niche Development, Regime Stabilization, and Positive and Negative Security

Given the fact Norwegian society is highly electrified and hydropower has been a long-established energy source, interest in new niche development in renewables has been lower in Norway than in the other case countries. However, alongside solar power and energy storage, offshore wind power began to attract more attention since 2022. Nonetheless, it could perhaps still be regarded a socio-technical niche in Norway.

For the same reason, and due to the hydrocarbon export sector being so important for the Norwegian economy (Andersen and Guldbrandsen, 2020), there has been very little talk of regime destabilization among politicians and energy businesses. The petroleum sector pursues decarbonization and not phaseout (Afewerki and Karlsen, 2022). There is some resistance to further plans for exploration by residents and fishers in certain areas (Korsnes et al., 2023). The latest policy from 2022 talks about the continuation and further development of the oil and gas sector – albeit a connected development is that of CCS. The latest IEA energy policy review of Norway also brought forward the question of critical materials and Norway's plan to begin producing them (IEA, 2022). Interestingly, this topic did not come up in the expert interviews in either phase. Therefore, the security issues discussed in this section are associated with stabilizing the existing regime and its security connections.

This section starts with the most dominant issue, that is, the economic and geopolitical security of Norwegian fossil fuel supply, which also serves as grounds to keep stabilizing this regime. Subsequently, the dominant electricity regime, namely hydropower and its security questions, are discussed. Finally, the section will end with the expanding wind power sector and the increasing importance of securing critical energy infrastructure.

### 7.5.1 Economic Security, Oil, and the Energy Transition

As this chapter has described, oil and gas production have provided substantial income for the Norwegian state, with the state being also a significant owner of the production capacity. The sector is also a sizable employer (Andersen and Guldbrandsen, 2020). A specific sovereign fund was created in the 1990s, which aims to invest this government money wisely in a way that benefits current and future Norwegians, that is, the whole of society: "Emphasis is placed on smoothing out economic fluctuations to contribute to sound capacity utilisation and low unemployment. The framework aims to preserve the real value of the fund for the benefit of future generations. It also isolates the budget from short-term fluctuations in petroleum revenue and leaves space for fiscal policy to counteract economic downturns" (IEA, 2022, p. 17). The way in which the fund has been created reflects elements of positive security, while it also in its own way contributes to increasing climate security risks. The oil fund is argued to be the largest sovereign wealth fund in the world, making up about one third of the Norwegian state budget. Therefore, oil production has brought economic security for Norway alongside global (geo)political influence.

## 7.5 Niche Development and Regime Stabilization

The economic security of Norway is so inherently tied to the oil and gas industry which is a massive problem. And it's a problem that the Norwegian government doesn't really seem to take in. And to have a society that is so dependent on one type of industry and high oil price. And we don't have to be, right? (Researcher, 2020)

To put it very bluntly and very simply is that Norway punches above its weight. It has gained some advantages both economically and politically, globally. It's simply because it has had lots of income. (Civil servant, 2020)

The political mindset has been much more about how to protect the income streams from these industries and no concern that we will run out of power or energy. (Business actor, 2021)

Yet some argue that this wealth has also enabled positive environmental developments, such as investments into environmentally beneficial solutions in Norway and abroad. The Norwegian government has made efforts to balance fossil fuel production and the climate and environment. Yet, it is clear that the broader global efforts to lower carbon dioxide emissions bring uncertainty to the Norwegian oil sector. This was especially the case during 2020–2021, although the substantial demand for Norwegian fossil fuels since 2022 has somewhat lowered these concerns.

If the demand side were suddenly taken away, it would be a big risk to unemployment and it would definitely reduce income to Norway, so that aspect of security is present. (Politician, 2021)

The fossil fuel sector is also tied to the more traditional protection of energy infrastructure against physical attacks. For instance, around 2010, the risk of terrorist attacks on Norwegian oil infrastructure was a security concern and military protection and surveillance are now used to protect oil and gas production.

More broadly, and especially since 2022, the Norwegian fossil fuel production has also been argued to be important for European energy security and global stability: "As a reputable and reliable producer, Norway has played a stabilising role in the world's oil and gas supply, particularly in meeting European demand" (IEA, 2022, p. 137). Norway is a stable insider partner for the EU with respect to fossil fuel supply, as a politician interviewed remarked. Recent strategies emphasize the continued development of the fossil fuel sector:

As the energy crunch struck, there was a sense of anything goes in Europe. Nuclear, coal, you name it. Any kind of energy to compensate for the loss of Russian imports. I believe that the right assessment is that the energy transition has suffered a temporary setback. Even if some people in Brussels and Berlin and wherever will contest that assessment. And I emphasize temporary, because there is little doubt about the direction of travel in the EU longer term. This of course creates a dilemma for a producer like Norway. (Business actor, 2023)

Today, it appears that the strong security arguments behind Norway's oil and gas production support its continuation for the foreseeable future. That is, arguments for the support of economic and geopolitical security of Norway as well as European energy security currently prevail. Hence, regime decline is not evident in terms of disruptions to skills and assets, unlearning, or deinstitutionalization.

### 7.5.2 Security of Hydropower

As noted, hydropower provides most of Norway's electricity production and, therefore, protecting this against intentional and unintentional disruptions is important. Annual precipitation levels influence the availability of hydropower. However, it can be stored, so, it is not affected by weather as much as wind and solar power are. Yet, changes in annual precipitation levels, affecting the amount of hydropower generation in Norway, mean that Norway is somewhat dependent on its neighboring countries for electricity supply during dry years. However, during the 1995–2020 period, Norway has maintained a power surplus for seventeen out of these twenty-five years (Hansen and Moe, 2022). According to the interviewed experts, hydropower infrastructure is not particularly vulnerable to external disruptions or attacks. One of the experts stated that generally hydropower has a very limited military component:

I think most of our large hydropower plants are inside the mountains. I mean there is maybe a one-kilometer tunnel that you have to drive through, so they are, not easy to access, from the outside. (Civil servant, 2021)

Nonetheless, an event in 2019 showed that the operation of hydropower companies can, for instance, be substantially disrupted by cyberattacks. In March 2019, a Norwegian renewable energy and aluminum manufacturer Norsk Hydro suffered a ransomware attack. The attack on the company, a producer of energy to 900,000 homes, encrypted key areas of the company's IT network on over 3,000 servers and locked everyone out (Austin, 2021). Norsk Hydro did not pay the attacker, which meant stalling production in all manufacturing facilities, because the company needed to shut down all access to the network and stick to manual operation of its critical systems for several weeks – costing approximately $70 million (Austin, 2021). Also, as mentioned, stormy weather in 2023 revealed the vulnerability of hydropower dams to extreme weather events, such as flooding.

Indeed, hydropower dam safety and preparedness are some of the few physical security issues mentioned in the expert interviews. Around 500 dams and reservoirs are categorized as having the highest risk of effect if they were to collapse. Increased securing of the dams to withstand larger attacks has driven up the costs of the hydropower sector in relation to wind power:

If one of these dams is destroyed it will give the most damaging effect on the society of all the installations we have in Norway. They are the most dangerous installations we have, because they are so huge and there's so much water behind them. (Civil servant, 2021)

## 7.5 Niche Development and Regime Stabilization

Some experts saw tensions in the security regulation concerning the dams, holding the view that civil servants were being overly cautious:

> But if you look in the area of security policy, military issues and energy, I've only noted one area where we have had significant conflicts or friction ... We believe that those risk assessments are not based on best science in the area and that they are very much higher than in other comparable sectors for example oil and gas or industry or let's say road infrastructure. And that they are very high compared to other countries that also have hydropower. (Business actor, 2021)

Such tensions have perhaps been alleviated since 2020–2021. The Nord Stream pipeline explosions in 2022 and the destruction of the Nova Kakhovka Dam in Ukraine in 2023 showed in real terms how energy installations can be targets of physical attacks and the substantial potential harm that may be caused. These events have emphasized the importance of securing critical infrastructure (see Section 7.5.4), that is, increasing the "negative security" orientation of the energy regime.

### 7.5.3 Tensions around Wind Power Expansion

Wind power expansion has been contentious in Norway, especially since the early 2000s (Korsnes et al., 2023), although is not directly linked to security. Due to the fact the Norway–Russian border is located at quite some distance from most of Norway's population, the operation of air surveillance radars was not seen as a problem to the same extent as in Estonia and Finland:

> For some of these windfarms that are located in close proximity to where we have military installations, that is an issue. But it's few and far between still. So, it's not something that generates lots of attention. (Civil servant, 2020)

Opposition to new wind power has been based on local environmental effects combined with the little added value perceived by those in opposition. One of the opposing parties has been the Center Party, which has voiced explicit concerns that Norwegian wind power is used to supply mainland Europe and paid for by Norwegian residents, industry, and the environment – something that has been labelled as a resource nationalist view (Hansen and Moe, 2022). This has created societal tensions, but not to such a degree that the many experts interviewed saw it as a question of internal security. Nevertheless, an ad hoc antiwind power organization, Motvind Norge, was established in 2019. This group was successful in halting all NVE-approved new onshore projects until April 2022 (Korsnes et al., 2023). There have also been other groups that have used extreme language and compared wind power to wartime occupation, treason, and the Nazi occupation (Hansen and Moe, 2022). The protests and opposition resulted in a freeze of new wind power permissions from 2019 until 2022, when the European energy crisis somewhat changed the outlook on wind power in Norway. Further, some experts interviewed mentioned the intention by some to physically damage new wind power development sites:

There have been extreme protests against building of windmills in Norway over the last two years. From being a symbol of progress towards an emissions-free energy system and something that has actually been stimulated through tax breaks and state aid, it has turned to be the symbol of destroying Norwegian nature. People don't want them anymore and that's been really a very difficult situation … So, there are no new wind parks being established after those that have already had a concession. And those that are being built are having large protests. (Business actor, 2021)

The White Paper issued in 2020 proposed changes that involved increasing local and regional involvement in onshore wind power projects and ensuring that environmental matters are taken into account (IEA, 2022), that is, improving the positive security perspective. In 2022, onshore wind development continued in cases where the local authority agreed with the state government.

### 7.5.4 Security of Critical Infrastructure and Supply since 2022

Security of supply was very rarely discussed by the interviewed experts in Norway prior to 2022. When it was mentioned, it was mainly in the context of electricity (as opposed to the defence sector security of supply, which pertained to oil and petroleum products). Only two out of fifteen interviewees brought this up during 2020–2021, and one noted that

There's no Norwegian politician, for many decades, who has woken up being afraid that there was no power. That's simply not, it's not our problem. Apart from technical collapse there's clearly no shortage. It's more price-related and that debate has been surprisingly fierce in my view. (Civil servant, 2020)

It is, however, important to note that the security of Norwegian energy infrastructure is also of importance to other countries. The Norwegian fossil fuel infrastructure in the North Sea is sizeable, with around 9,000 kilometers of pipelines. For instance, a disruption to the gas pipelines between Norway and rest of Europe would lead to substantial harm to countries at the receiving end (Godzimirski, 2022). Moreover, the events of 2022 led to a realization that, although Norway has frequently been the energy exporter, there is mutual reliance on other countries, which is also beneficial for Norway during times when hydropower resources are smaller. Despite this recognition, in Norway the energy crisis did not materialize to the same extent as in rest of Europe and, as an interviewee put it, there is not the same sense of urgency in Norway as elsewhere in Europe.

Since 2022, focus on critical infrastructure in Norway has markedly increased and security zones around energy installations have been broadened. The interviewees reported an increasing number of military capabilities assigned to the North Sea, highlighting the rising negative security focus on energy infrastructure. This has been a result of the war conducted by Russia in Ukraine as well as the explosions of Nord Stream pipelines.

On the security side, it was a wake-up call, when we had the explosions close to Nord Stream 1 and 2. In the lead-up to those explosions, there were observations in Norway on drones close to our onshore installations and also offshore installations. (Civil servant, 2023)

Both the government and the company have been extremely focused on doing what we can to prevent or mitigate attacks in the cyber domain or indeed physical attacks on sea or land installations or against cables on the seabed. We have seen, this is public knowledge, we have seen the home guard, a branch of the armed forces, patrolling our main onshore installations, and also armed forces have contributed to the patrolling of the Norwegian continental shelf. (Business actor, 2023)

Another security implication, associated with positive security and people's experiences, links to increasing electricity prices. Higher prices, to which Norway's residents had been unaccustomed, made energy more politicized than before. The political discussions also covered international powerline connections, where some extreme voices argued that Norway should halt its international electricity supply:

Now energy policy is left, right and center in all political discussion, and it didn't used to be. This [energy] was something we were taking for granted or for given before, but now we see that this is actually not adding up [to] the demands. (Civil servant, 2023)

Increased demand has put us in a position where Norway occasionally has had the highest energy prices in Europe, which is a significant political challenge because the Norwegian population really doesn't understand how this can be. (Civil servant, 2023)

It became very nationalistic, in the beginning, [i.e.,] we need to make sure that we have the power that we need and make sure that we can keep it for ourselves. Also, on the political side, there were those types of discussions. (Energy business, 2023)

Although the Norwegian government has compensated consumers for high electricity prices, questions about sovereignty and control over electricity exports have been increasing for some time. In particular, exports based on renewable energy – especially wind power – have raised a resource nationalist discourse based on claims that such exports weaken Norway's energy sovereignty (Hansen and Moe, 2022). It is true that, unlike in the other case countries of this book, Norway's sovereignty is not directly improved by the renewable energy expansion. However, the degree to which such expansion actually leads to adverse sovereignty depends on the perspective taken. Making a distinction between actual and perceived sovereignty, Hansen and Moe (2022) pointed out that the resource nationalist discourse, which claims renewable energy exports have adverse impacts on Norwegian sovereignty, is not new but has simply received increasing attention in the last few years. They say that this discourse is not targeted at Norway's operation in the Nord Pool power exchange but is related to the country's relationship with the EU. Yet, the counterargument to the sovereignty concern is that, without export, the Norwegian electricity sector is not viable when prices fall too low due to excess supply (Hansen and Moe, 2022) – not to mention the beneficial effects on European stability of Norwegian electricity exports.

Some interviewees argued that a degree of securitization of energy policy has happened in terms of these questions on control and sovereignty, in addition to the increased perceptions of energy infrastructure as a critical and potential target for military attacks. A further example is bringing Equinor under the Security Act. None of these issues were regarded by the interviewees as examples of a high degree of securitization, though. When taking the view presented in the political economy of energy research on securitization, securitization has happened in Norwegian energy policy to the extent that energy security as a term has become part of the vocabulary in Norway, which it wasn't before. Further support for some degree of securitization are the exceptional measures taken during 2022–2023 as part of Norway's energy policy: the more visible military presence safeguarding the critical energy infrastructure and accelerated support for wind power. Some have remarked that higher fossil fuel prices are an additional incentive for the defence sector to implement energy-efficient measures:

> It's definitely been more securitized on [the] political level, and after the explosion of the Nord Stream 1, there [have] been measures and increased attention of the need to make sure that you have both physical and also technical security around the energy grid. (Energy business, 2023)

Yet these measures and changes in rhetoric would not be sufficient to meet the security studies' definition of securitization.

## 7.6 Concluding Remarks

Norway presents as a very different country case study compared to Estonia and Finland in the previous chapters. In sustainability transition terms, one can say that regime stabilization of the oil and gas sector is taking place instead of regime destabilization. Wind power niche expansion has been rather modest and contentious, especially since 2019, although it has been gaining new traction since 2022. Moreover, energy security only appeared in the vocabulary after 2022.

The importance of the exporting hydrocarbon sector for the Norwegian economy can be framed both in terms of positive and negative security. There is much

## 7.6 Concluding Remarks

long-term positive security associated with oil and gas via the revenues generated, which have been used to improve the social security and living standards of Norwegian residents. This has enabled a sense of "freedom from insecurity" and community in Norway (cf. Booth, 2007). In addition, there is negative security associated with possible future regime decline, whereby Norway would lose the economic and geopolitical security its fossil fuel reserves offer. The petroleum sector export and revenue have given Norway more geopolitical power and leverage than a country this size would otherwise have and losing this leverage might be seen as a kind of a security threat. Nevertheless, there is also implicit negative security associated with the continuation and further development of the fossil fuel regime. As the negative and disruptive impacts of climate change via adverse weather events and melting permafrost become increasingly evident, this sector – supplying circa 3 percent of world's oil supply – contributes to increasing climate security risks in the future.

Expectations for future development are divided – or perhaps held together by dissonance – around the continuation of the fossil fuel regime and the partial switch of the electricity regime to wind power, although the latter has been much weaker than the former (see Figure 7.5). Explicit interconnections between energy transitions and security questions were missing before 2022 – this interface was exemplified mostly by incoherent policymaking and a lack of policy coordination mechanisms. There has been a marginal debate whereby some actors have associated negative security, via adverse sovereignty, with the sustainable energy transition (Hansen and Moe, 2022). This is linked to resistance to the transition, based

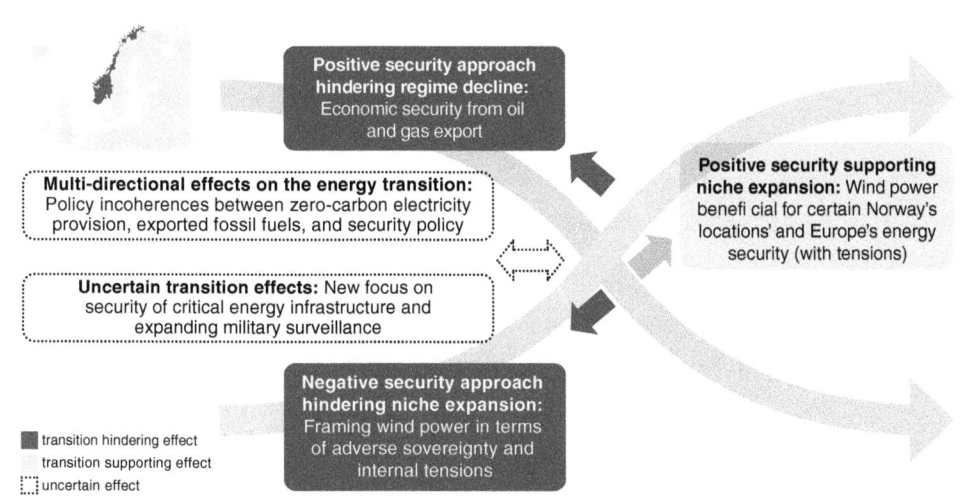

Figure 7.5 Key energy security aspects and their transition impacts in Norway, 2006–2023.
Source: Kivimaa, Finnish Environment Institute, 2023.

on opposition to wind power and partly on higher electricity prices due to international electricity interconnections. This has seen aggressive use of language yet, so far, few physical security implications. Nevertheless, one could argue that the negative and polarized discourse reduces the overall positive security of Norwegian society.

Finally, a clear repoliticization of Norwegian energy policy took place in 2022, which also has dealt with these questions of energy sovereignty and energy security that have become part of Norway's energy policy vocabulary. A strong degree of securitization is not evident, but there have been breaks from previous energy political practices – evidenced by new support for offshore wind power and visible military protection of critical energy infrastructure.

# 8

# Scotland

*From Oil to Wind under a Devolved Government and New Pressures for UK Energy Security*

The last country case in this book deals with two dimensions. First, it looks at the UK as a whole and how its energy policy interacts with security and defence policy. Second, it zooms into Scotland as a somewhat comparable country to the other case countries in terms of population size (5.5 million inhabitants), its Nordic location, its pursuit of a zero-emission society, and its rather liberal values. As Scotland's devolved powers do not include energy policy as such, nor security and defence policy, it mainly addresses energy questions via its powers over energy efficiency and land-use planning policies.

UK energy policy is an interesting mix of relatively strong support for fossil fuels coupled with ambitious long-term climate targets via the world's first Climate Change Act, which came into force in 2008. The British coal industry has a long history and began destabilizing a hundred years ago. While coal mines began to close on an accelerated basis in the 1960s, coal sales to the electricity sector continued to expand until the 1980s. A "dash for gas" in the 1990s meant replacing coal as a source of heating fuel for buildings and was a start in reducing the carbon dioxide emissions of the energy sector. However, it was a specific policy instrument, the Carbon Price Floor, in 2013, that contributed to a rapid reduction in coal generation, from 39 percent in 2012 to only 2 percent in 2021 (DESNZ, 2023). The British coal phaseout, however, led merely to coal being replaced with natural gas for heating, which also has rather substantial greenhouse gas emissions. The development of the UK energy mix shows a considerable decline in coal use while the use of petroleum products and natural gas has remained relatively constant (Figure 8.1). In the overall energy mix, the share of renewable energy is rather modest. The extent of the gas network makes achieving heating reform by moving to nonfossil energy sources difficult. The electricity sector has, however, progressed with decarbonization, with 65 percent of electricity produced from renewable energy in Scotland and 38 percent in England and Wales.

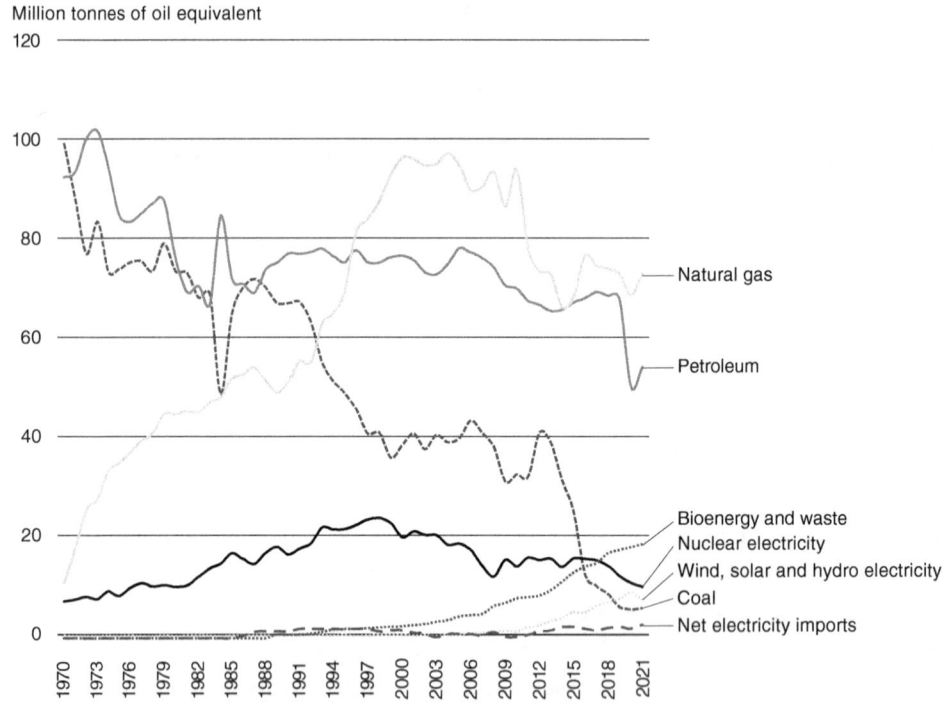

Figure 8.1 Inland consumption of primary fuels and equivalents for energy use, million tons of oil equivalent.
Source: DESNZ (2023).

A kind of rejuvenation of "noncoal fossil fuel" policy happened in 2007, when hydraulic fracking for shale gas was added to the agendas of resource companies and the government alike. Some scholars argue that fracking represents a reproduction of the fossil fuel hegemony justified on the basis of issues such as employment and energy security (Nyberg et al., 2018). The different UK governments' stances on hydraulic fracking of gas and oil have fluctuated from support to bans. In the end, fracking has never taken off properly and has ended for now. Overall, the past fifteen years or so have seen the contradictory parallel tracks of aiming to decarbonize and keeping hold of the fossil economy – somewhat similarly to the Norwegian case.

Scotland has been – and is – important for the UK energy and security sectors. Aberdeen (and the surrounding area) is the center of the British oil and gas industry. It has been described as a nexus of Europe's fossil fuel industry and is the headquarters for many companies (Adams and Mueller-Hirth, 2021). Further, England's electricity demand exceeds its electricity generation, so it needs transfers of electricity from Scotland, Wales, and Continental Europe. Scotland produced 57 percent of the UK's renewable electricity in 2021 (DESNZ, 2023).

Scotland also has the lowest share of fossil fuel-based electricity generation in the UK and, while it is the hub of fossil fuel production, it uses little of the fossil fuel it produces (akin to Norway). In 2020, 97 percent of Scotland's electricity generation was from renewable energy (Scottish Renewables, 2023). This means, in practice, a fully renewable energy-based power supply. Scotland's wind power capacity in June 2022 was 13.3 gigawatts (GW) and, at that time, a further 16.7 GW of wind power was under construction or planned (Scottish Government, 2022).

One of the key future concerns for Scotland is the substantial decline of oil and gas production from the North Sea. In 2019, the oil and gas industry supported 57,000 direct and indirect high-value jobs and accounted for 9 percent of Scotland's GDP. While the sustainable energy transition will create new jobs, the employment capabilities of the fossil fuel-exporting industry are hard to replace, creating potential socioeconomic insecurity for Scotland. Therefore, support for wind and hydrogen energy is argued to require significant early investment and policy support (Earnst & Young, 2023). The Scottish government wants the fastest possible just transition of the oil and gas sector, by investing, for example, in reskilling fossil fuel workers in the renewables industry and hydrogen sector (Scottish Government, 2023).

Scotland is also an important location with respect to the UK defence system. The country's Faslane and Coulport naval bases are where UK military nuclear submarines are located. This is despite the opposition of many Scots, and especially the Scottish National Party (SNP), to the use of nuclear technology for energy and security purposes.

Finally, Scotland is an interesting case due to its attention on just transitions. The Just Transition Commission was established in 2018 as one of the first concrete just transition developments in the world.

This chapter differs from the other country chapters in that it examines both UK policies as a whole and then zooms into the specificities of Scotland, which has some devolved powers. The chapter describes the key context, that is, the energy and security regimes. It then continues with subsections, drawing on Chapter 4, namely the perceptions of Russia as a landscape pressure for energy transitions; policy coherence and interplay between energy and security regimes including the level of securitization; and, finally, positive and negative security related to niche development and regime (de)stabilization. The data used to inform the analysis comprises energy- and security-related government strategies published since 2006 and two rounds of interviews with energy and security experts, the first between November 2020 and April 2021, and the second in January 2023. The data analysis has been complemented by literature sources.

## 8.1 Energy Regime

### 8.1.1 The United Kingdom

The UK energy regime has a long history, during which the sector has moved from nationalization to privatization. The post-World War II period experienced the nationalization of the energy sector until the 1970s. Prime Minister Thatcher's government began strongly privatizing the energy sector with the hope of taking advantage of low international energy prices and technological innovation (Bolton, 2021). However, the related closure of many coal mines led to mass labor unrest and national strikes in the mid-1980s. Privatization led to the liberalization of the energy sector in the 1990s – as in many other countries – where production and supply were unbundled and competition opened more fully. In 2000s, decarbonization of the energy sector also became a key part of energy policy.

A Labour Party government in 2008 created the Department for Energy and Climate Change (DECC), subsuming energy-related responsibilities from previous economic and environmental ministries. This was linked to the more ambitious climate policy of the government at the time, with the introduction of the Climate Change Act in 2008, and a White Paper on "Energy and Climate Change" in 2009. However, less than a decade later, in 2016, the Conservative government then in power emphasized energy innovation more than decarbonization of the energy sector, and created the new Department of Business, Energy, and Industrial Strategy (BEIS). This change stressed the revival of industrial (energy) policy (Johnstone et al., 2021) and was preceded by the removal of many energy transition policies supportive of renewable energy and energy efficiency (Kern et al., 2017).

The latest administrative change occurred in 2023 with the establishment of the Department of Energy Security and Net Zero (DESNZ), which reflected the changes in the energy landscape caused by the Russian war in Ukraine and the European energy crisis of 2022 (see Section 8.3). DESNZ is oriented, for instance, toward building energy efficiency. However, generally the lack of attention paid by UK governments to energy efficiency as something that improves energy security has often been criticized. The organizational change may also mean a repoliticization of previously depoliticized energy policy but this is not guaranteed. Indeed, the UK has a long history of a market-led energy policy paradigm, which has become deeply entrenched and may be difficult to overcome even in these changed circumstances (see Lockwood et al., 2022).

While generally, in the UK dialogue on climate and energy policy, the importance of the EU has been downplayed, academic research has pointed to the contrary. The EU acted as a significant supranational institution for UK climate policy and in particular influenced its renewable energy policies, which would have been less ambitious without EU influence, according to estimates (Lockwood, 2021). The same applies for the influence of the EU on building energy efficiency policies (Kern et al., 2017). Therefore, Brexit – the departure of the UK from the EU in 2020 – has substantially changed the energy policy setting in Scotland and the rest of the UK.

The energy regime in the UK can be described as being under centralized government power – with relatively frequent shifts in the composition of ministries – with the "Big Six" energy companies (British Gas, EDF Energy, E.ON, npower, Scottish Power, and SSE/OVO), or more recently the "Big Five" after E.ON acquired npower, having a great deal of influence. The ownership of wind energy has also been concentrated within the Big Six (Lockwood et al., 2022). With the advancement of decarbonized power supply, the number of suppliers has increased, resulting in around eight with over 5 percent of market share each and smaller suppliers with circa 8 percent of market share in total in the 2020s (Ofgem, 2023). Scottish Power and SSE are the two overall largest companies in Scotland (Wilson, 2022), showing the importance of energy for the Scottish economy.

The Office of Gas and Electricity Markets (Ofgem) is an important agency in the energy administration, operating under the supervision of the ministry in charge of energy. Its role is to regulate companies that operate gas and electricity networks, decide on price controls, and primarily to protect the interests of consumers, being an important actor in liberalized energy markets. While tasked also with overseeing the fulfilment of environmental considerations, Ofgem was mostly focused on prices and competition until the late 2000s (Pearson and Watson, 2012). It has been characterized as having a significant degree of regulatory independence, being a lead actor in some areas of energy policy, with this power somewhat slowing down its orientation toward decarbonization (Lockwood et al., 2017). The Energy White Paper from 2020, "Powering Our Net-Zero Future," stipulated that Ofgem also needs to have a role in advancing the zero-carbon transition: "Subject to Parliamentary approval, the Strategy and Policy Statement will require the Secretary of State and Ofgem to carry out their regulatory functions in a manner which is consistent with securing the government's policy outcomes, including delivering a net zero energy system while ensuring secure supplies at lowest cost for consumers" (HM Government, 2020, p. 86). With regard to the energy transition, Ofgem has earlier recognized as problems the limited role of consumers and lack of consumer-oriented business models (Johnstone and Kivimaa 2018). The experience with Ofgem shows that state-mandated organizations with a great deal of independence may in part slow down the energy transition, while such independence could alternatively be used to exceed the decarbonization goals of the state.

Two independent advisory groups operate under parliament with opportunities to critique and comment on government climate and energy policies: the Committee on Climate Change (CCC) and the National Infrastructure Commission (NIC). The CCC was established in 2009 based on the Climate Change Act. It sets carbon budgets that the UK government must meet and which the CCC evaluates annually. The CCC comprises the Climate Change Committee with a chief executive and six academic members, the Adaptation Committee with a chair and five academic members, and the secretariat. The CCC has been regarded as successful in safeguarding long-term policy continuity; its analytical orientation and political awareness has, however, depended on who holds the chairmanship (Fankhouser et al., 2018). The NIC includes two chairs, five commissioners, and the secretariat, with the task of providing impartial advice on infrastructure to the government.

The National Grid has been responsible for maintaining the electricity transmission and gas networks and for security of supply since its establishment in 1935. The company owns one of Britain's three onshore transmission networks and four electricity distribution networks, and the electricity system operator. However, the Energy Act of 2023 stimulated that the latter is replaced with a government-owned independent public corporation National Energy System Operator (NESO) and a Future System Operator, to become operational in summer 2024. The National Grid is a fully privately owned transmission system operator (TSO), unlike other European countries' TSOs, and one of the largest investor-owned utilities in the world with a significant share in foreign ownership (Yates, 2022). This means that, before the establishment of NESO in 2024, responsibility for supply security has been allocated to the profit-seeking private sector and that the owners of the National Grid may not have been so keen on the kind of transmission network investments that the zero-carbon transition requires. This highlights a feature of UK energy governance whereby power has been partially delegated to the energy industry (Lockwood et al., 2022), associated with the depoliticization of energy governance, that is, a lack of political scrutiny (Kuzemko, 2014). The move to a government-owned electricity system operator shifts this setting to similar direction as the other case countries. However, as in many countries, in the UK distribution network operators are often privately owned, and this has implications for network development.

### *8.1.2 Scotland*

The Scottish government's climate and energy policy is more ambitious than the UK government's, but it does not have substantial policymaking power, with most resources at the UK government level (Lockwood, 2021). The principal way in which the Scottish government can influence energy production and use are the land-planning policy and energy efficiency measures. In essence, "key aspects of energy policy are 'executively devolved,' including control over major energy consent and planning, and operational control over aspects of market support" (Cowell et al., 2017, p. 173). Yet the fast growth of renewable energy in Scotland, an almost 2.5-fold increase between 2012 and 2022 (Scottish Government, 2022), has increased the political negotiation power of the Scottish government with Westminster (Cowell et al., 2017).

As a former BEIS civil servant pointed out:

> So, while energy policy broadly defined is a reserved power for the UK government, even across energy policy there's a huge amount of consultation and engagement that happens between the Whitehall and Scottish government and various parts of Scottish sub-national government et cetera. These things are never really done genuinely in isolation.

Key themes in Scottish energy policy have been the promotion of renewable energy, the energy efficiency of households, and opposition to new nuclear power stations since about 2010. Increasing attention has been paid to energy poverty and energy justice (Santos Ayllón and Jenkins, 2023). Energy security has received relatively little attention and has mostly been perceived, prior to 2022, via a "markets will deliver" approach, as in the rest of the UK, with no specific Scottish policy on this. An energy expert argued in 2021 that too little attention was paid to the development of energy storage and smart grids as potential facilitators of energy security, with overreliance on the National Grid to deliver.

Scotland has its own Climate Change Act, issued in 2019, and has a net-zero target to reach by 2045. Some interviewees argued that the Scottish Climate Change Act has high targets but has inspired little concrete action. In late 2022, the CCC strongly criticized the lack of concrete plans and insufficient policy progress toward the Scottish climate targets and revealed substantial off-track developments in many areas (CCC, 2022a). Unlike the UK government, the Scottish government has opposed both nuclear power and the fracking of shale gas.

A draft "Energy Strategy and Just Transition Plan" was issued by the Scottish government in early 2023. This plan aimed to more than double wind power production, with increased contributions from solar, hydro, and marine energy, and outlined the establishment of a new energy agency – Heat and Energy Efficiency Scotland – alongside emphasizing energy security much more than before (Scottish Government, 2023). Further, it noted that

> The Scottish Government is clear that unlimited extraction of fossil fuels is not consistent with our climate obligations. It is also clear that unlimited extraction, even if the North Sea was not a declining resource ..., is not the right solution to the energy price crisis that people across Scotland are facing or to meeting our energy security needs. (Scottish Government, 2023, p. 97)

A small number of actors, including Scottish Power, dominate Scottish energy policymaking. An expert from a think tank argued in 2021 that the Scottish government is influenced by lobbyists representing large companies and it is headed by former industry representatives. This view was shared by an academic who also perceived that the Energy Saving Trust had a great deal of influence, although environmental nongovernmental organizations (NGOs) also played an important role in Scottish policymaking.

The dominance of economic and free market-based energy policy in the UK has been obvious. This originated at least as far back as the 1970s, with emphasis on economics, "rational choice," and free market, but there have been some periods that were more regulation-orientated, especially during 2006–2008 when the Climate Change Act was enacted (Kivimaa and Martiskainen, 2018). This contrasts with the somewhat more sociodemocratic approach of many Scots –which takes account of the natural environment and the welfare of the poor – even though some contrary views also exist.

My experience is simply: the UK is hard-core free market in its approach to energy. The Scottish government has no ideological approach to energy and has been content to let industry lead. The outcome is that the political perspective is purely about "public communications" – presenting Scotland as a "green powerhouse" or a "world leader in renewables." There seems next to no interest politically in how that is actually developed. (Think tank, 2021)

Nevertheless, Scotland was one of the first countries to form a just transitions body. The Just Transition Commission, set up by the Scottish government, was influenced by a coalition on just transition formed by NGOs and the Scottish Trade Union Congress and established to monitor and counsel on government climate policy according to just transition principles (Santos Ayllón and Jenkins, 2023). An interviewee noted, however, that it had little political or business influence. Figure 8.2 summarizes the key aspects of UK and Scottish energy policy.

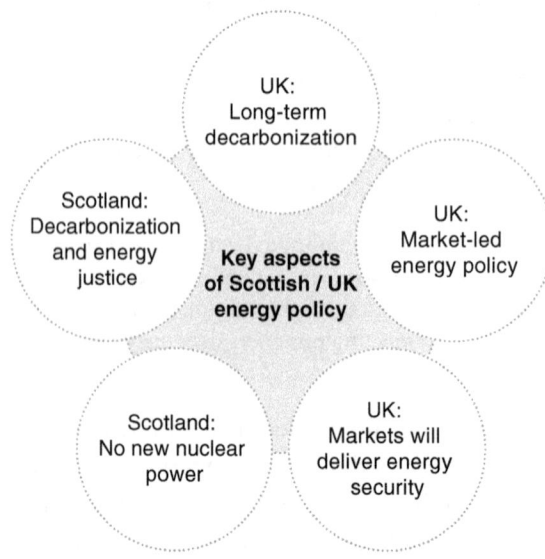

Figure 8.2 Key aspects of UK and Scottish energy policy.

## 8.2 Security Regime

Britain is an island state so its defence and security policy are related to the control of waters and maintaining free movement of trade, but also to the centrality of the North Atlantic Treaty Organization (NATO) and close cooperation with the US and France (Dorfman, 2017). The US has been the global hegemon with whom the UK has closely built its security and defence regime (see Regilme and Hartmann, 2019). Perhaps linking to the UK's past global power and international security cooperation, the UK has military bases with a wide global reach. They are located in countries in different parts of the world, such as Canada, Belize, Kenya, and Iraq. The UK also wields soft power via extensive diplomatic efforts and relations globally, albeit with some loss of power after Brexit. These factors were apparent in expert interviews in terms of connections to climate and energy questions.

The first "National Security Strategy" was published in 2008 by the Cabinet Office. It was criticized for a lack of consultation with other departments and, hence, was revised as early as 2009, this time employing cross-government consultation (Dorfman, 2017). The strategy was followed by the establishment of the National Security Council and the post of National Security Adviser in 2010. Since then, each parliament has been expected to produce a parliamentary defence and security review, titled, since 2019, "Integrated Review of Security, Defence, Development, and Foreign Policy." The integrated reviews produced by UK governments have increasingly crosscut several policy domains. The 2023 integrated review stated that: "[t]he most pressing national security and foreign policy priority in the short-to-medium term is to address the threat posed by Russia to European security" and emphasized increases in defence spending (HM Government, 2023, p. 11). It also emphasized broader, nonmilitary security aspects, connecting, for example, to the new Critical Minerals Strategy and the new Semiconductor Strategy. This illustrates the increased importance to secure the supply and promote the science-and-technology development of these materials and components, not only for energy but also for other digitalizing sectors.

Defence policy has as its objective to protect people, to stop conflicts, and to be prepared for potential battles (Ministry of Defence, 2019). Historically, delivering access to oil and gas and securing international supply chains has also been one of the tasks of the UK military sector. In addition, during the current millennium, climate change mitigation has become a part of UK defence policy (Depledge, 2023). British defence policy has also been interested in the Arctic region and the High North, despite the UK not being one of the Arctic states. This concern intertwines with climate change, because defence officials have been concerned about

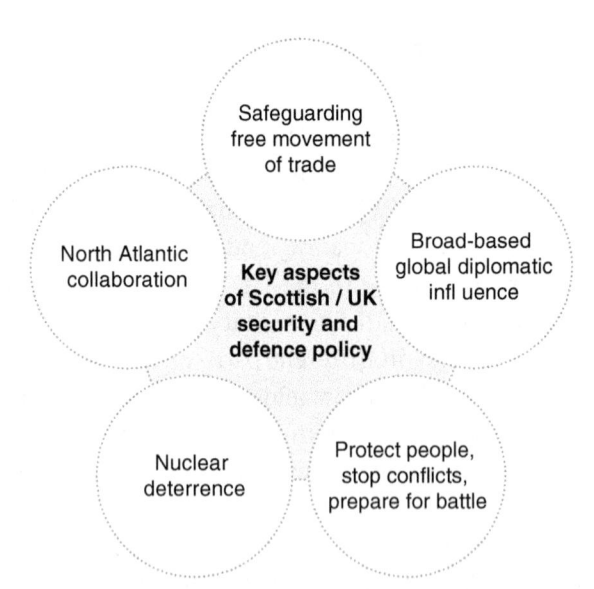

Figure 8.3  Key aspects of UK security and defence policy.

- Safeguarding free movement of trade
- North Atlantic collaboration
- Broad-based global diplomatic influence
- Nuclear deterrence
- Protect people, stop conflicts, prepare for battle

Key aspects of Scottish / UK security and defence policy

the implications of climate change on Arctic resources and trade routes becoming more available in response to sea-ice melting (Depledge et al., 2019). Figure 8.3 summarizes the key aspects of the UK security and defence policy.

Central public actors in security, defence, and international relations include the Ministry of Defence (MoD), the British Army, and the Foreign, Development and Commonwealth Office (FCDO). Scotland does not have devolved powers related to security and defence. Therefore, it is dependent on UK government agencies, not only for military defence, but also for cybersecurity (Neal, 2017). The UK's defence spending was 2.2 percent of GDP in 2021 (HM Government, 2021), having had a declining trend until 2022. The state possesses over 200 nuclear weapons, of which 120 are active, with 40 positioned at a time in four nuclear-powered submarines (Dorfman, 2021). These are located in Faslane and Coulport naval bases in Scotland, although Scotland does not have any say regarding the location of the nuclear weapons. A recent decision was also made by the UK MoD to build twenty-six new military vessels for the Royal Navy in Glasgow.

It is obvious that the political ambitions and worldviews in Scotland differ from that of the broader UK regarding security, defence, and foreign policy:

Well, I think it's relevant to energy because Scotland, there really isn't an interest in being a global power anymore. That's not a worldview that's held. There's no idea that, of Scotland's got to be in the UN Security Council with America and Russia. It's not an aim. And for Britain, it is a central aim. (Researcher, 2021)

The reports produced in connection to the 2014 referendum for Scottish independence emphasized a different approach to defence than that in the broader UK (Neal, 2017). An independent Scotland led by the SNP would prefer to be a part of NATO while refusing to hold nuclear weapons or implement nuclear deterrence measures (Ritchie, 2016). However, there are also those who see that Scotland would suffer from independence and, for instance, lose the defence dividend paid by the rest of the UK to Scotland (Fleming, 2021). Therefore, it would be difficult to depict what the security and defence policy of Scotland would look like if it existed.

## 8.3 Perceptions of Russia as a Landscape Pressure at the Intersection of Energy and Security

The landscape pressures in energy and policy documents, at the energy–security nexus, were similar to those presented in other countries. For instance, during 2006–2010, globally increasing competition for energy, coupled with some states using energy as a "hostile policy tool," was regarded as a security concern. Increased supply disruption risks were anticipated due to social unrest and corruption, while nuclear safety was also a specific landscape concern for Scotland. Russia was perceived as a landscape pressure via the Russia–EU gas dispute, and as applying energy as a political lever. New landscape pressures during 2010–2015 included Arctic developments, nuclear proliferation, and crowding of sea space with transport and offshore energy infrastructure. In 2015–2020, policy documents noted the pressures of climate change, pandemics, cyberattacks, and nuclear and chemical weapons (Kivimaa and Sivonen, 2021). There was very little discussion of Russia per se as a landscape pressure in the policy documents, even after the high media attention of 2006.

Nevertheless, Russia has been one of the countries of "landscape" concern for UK security and defence policies, as with the other case countries in this book. Due to the absence of a joint border, however, the perceptions of Russian pressure have, in this energy–security nexus, perhaps been less substantial. The changes in Russian energy policy from 2004 onward and the Russia–Ukraine gas dispute in 2006 "marked a reversal from the politically and ideologically significant processes of privatisation and liberalisation that Russia had initiated after the Cold War ... they represented the polar opposite of the free markets that UK policy makers, and other institutions, such as the International Energy Agency (IEA), had been so actively seeking to establish" (Kuzemko, 2014, p. 265).

Broadly, it seems that, prior to 2022, the influence of the "Russia risk" has been minimal on the UK and Scottish energy regimes. It was barely mentioned by the interviewed energy experts, compared to interviews concerning the other case countries. Some even expressed very relaxed attitudes to any concerns:

Russia has an economy that shouldn't bother us in the slightest – it is small and vulnerable. It has (as best as I can tell) little capacity to mess around with energy exports based on its economic interests alone. (Think tank, 2021)

I think the EU needs to get over its paranoia about Russia. I find this discussion of Nord Stream 2 … just, the amount of energy and effort wasted on it is unbelievable. I think at the end of the day, what you need is a functioning global gas market. (Researcher, 2021)

However, in hindsight, the latter person remarked two years later:

I think there was a degree of complacency because we only got 4 percent of our gas from Russia, and I think very quickly it became clear that we were exposed to Europe's dependency, and because of our reliance on gas being much greater than just about any EU country, with the possible exception of the Netherlands, we've been particularly hard hit by the high gas price. (Researcher, 2023)

There were select interviewees, however, principally those from the defence and foreign policy sectors, who have always considered the Russian state to be an energy security risk in political decision-making processes. This aspect has been used to legitimize nuclear power by the UK government and renewable energy by the UK and Scottish governments. While, initially, the Scottish decision-makers may have been more concerned than UK decision-makers, their perception became a consensus in 2022:

I think politically in Scotland, the SNP as a whole tends to be slightly more worried about Putin's Russia and the type of influence it exercises in the world. (Politician, 2021)

For some people the indirect links between Russian and UK energy systems were visible, while, for most, the links only became visible via the energy crisis ensuing from the war:

We see Russia, not necessarily a wholly positive player, our relationship with Russia is difficult at the moment. They will use their economic power, their influence, for their own benefit. We need to counteract that, and make sure our own supplies, as we have done, are diversified, but we are also conscious of the possibilities that, say, a move into Ukraine will have for global security. (Civil servant, 2021)

What this tells us is that, first, geopolitical concerns over Russia have not been particularly important for the Scottish or wider UK energy policy, as the country is geographically remote. Nevertheless, the energy transition has received some legitimacy from the indirect dependency of the UK on Russian energy flows. More broadly, however, it also shows the high level of market orientation of UK energy policy, which has been more about economic developments than about different dimensions of security. The events of 2022, however, showed that developments concerning Russia also influence the energy sector substantially, in this case in terms of significant energy price increases.

## 8.4 Policy Coherence and Interplay

### 8.4.1 Interaction between Energy and Security Issues

The relatively small importance of Russia, or geopolitics more generally, for Scottish and UK energy policy is likely to also have moderated the attention that UK energy policy and the energy administration have paid to horizontal coherence between energy policy and security and defence policies and to integrating security aspects into energy policy. Some argue that the discussion on security in the context of UK energy policy has been tightly limited to "energy security" and, even then, mostly on the kind of security that markets can deliver. Even the new "Energy Security Strategy," which was issued in 2022, had few concrete actions to create a more secure low-carbon energy system for the UK and still too little attention paid to energy demand reduction and energy efficiency. In the Westminster Research Forum on Energy Security, organized in October 2022, the chair remarked: "It is very clear we need new nuclear power."

There are, however, deeply rooted interconnections between the energy and security regimes in the UK. The production and expansion of oil benefited twentieth-century war efforts (Johnstone and McLeish, 2022). Further, oil has been connected to UK military efforts in contributing to maintaining stability in the Persian Gulf:

Presence in the Middle East was one of our military tasks we had to perform what we were directed by the government to do in order to ensure free flow of trade and other issues related to UK diplomacy and interaction ... It was principally one that fell to the maritime environment and if you're in the ship you're going to find yourself operating there as much as you going to find yourself operating in the Caribbean or Far East. (Former navy official, 2021)

Consequence is that we then have British warships and NATO ships in that region on anti-piracy missions, so there's a national security impact there and that also then brings increased instability to the supply lines. (Civil servant, 2021)

These energy–security connections relate more to energy in the external context rather than within the UK.

An analysis of policy strategy documents conducted in 2020 showed that, with regard to objectives, during 2006–2015 there was a rather high level of integration between energy and security policies, which was visible, for instance, through remarks related to energy efficiency, low-carbon technologies, security enhancements at critical energy sites, and Royal Navy ships protecting oil platforms (Kivimaa and Sivonen, 2021). However, from 2016, the policy documents paid much less attention to this integration, coinciding with the change of government in 2015 that also resulted in removing many policy instruments supporting low-carbon technologies and building energy efficiency. This means that energy policy was mostly made on economic- and market-based premises with declining interest in both decarbonization and security.

Regarding the latter, the UK energy sector has been governed with the idea that free markets, that is, balancing supply and demand, will deliver energy security. This is a result of a long-term depoliticization of energy policy and the placing of energy policy in technocratic contexts: "Arguably, the placing of elected representatives at a remove from active deliberation also resulted in lack of political capacity to engage with and understand energy and its relationship to wider societal goals, such as security" (Kuzemko, 2014, p. 262). There has also been a contrast between a coherent fossil fuel and security approach to safeguard international fossil fuel routes and the advancement of climate security via the energy transition. For instance, policy documents from the period 2011–2015 framed declining domestic fossil fuels production as a security risk, which conflicts with low-carbon energy policy (Kivimaa and Sivonen, 2021). This insight was confirmed in the expert interviews conducted in 2021 and again in 2023.

In energy policy, energy security was emphasized, particularly during the first decade of the 2000s, following the UK's increasing dependence on imported gas, while it was little discussed around 2020. Moreover, energy security has been understood as physical security and self-sufficiency in fuels. This thinking applies also to renewable energy in a sense. The reduced attention of energy policy strategies toward security of supply had practical implications, such as the closing down of Centrica's Rough gas storage site in 2018, only for it to be reopened after the 2022 events.

International electricity interconnections and the development of demand-side response and energy storage have been regarded as important means of energy security in a low-carbon energy system. Yet, the 2022 CCC progress report, which mentioned security over 100 times, criticized the government's energy security strategy for not employing demand-side measures that would benefit energy security. It also provided a specific policy recommendation related to improved coherence:

The Government's 2030 Strategic Framework should set out how the international climate and environment capability built up during the UK's COP26 Presidency will be resourced, maintained and further developed to enable delivery of international climate goals. Particular focus should be given to plans for coordination and consistency across departments and the embedding of dedicated climate experts in areas such as trade, security and foreign policy. (CCC, 2022, p. 40)

In addition, it warned about the risk of lock-in to new fossil fuel infrastructure, such as export and import terminals for liquefied natural gas. Such a risk is a real possibility given the low advancement of energy efficiency and heating infrastructure changes in the UK, combined with poor-quality building stock and increasing energy poverty. Hence, the dependence on natural gas is still high.

A related issue, which had already received some attention in 2021 and has since become much more prevalent, is the supply of critical materials – minerals and metals required for low-carbon energy system technologies and infrastructure as well as other digital devices. In addition, the supply of renewable energy technologies and components is a security concern. European countries have acknowledged the substantial role of China in the global trade-and-supply chains of critical materials and renewable energy technologies, and have been developing strategies to respond to this. In late 2022, the UK announced significant funding for battery research and innovation (BEIS, 2022), and the construction of its first lithium refinery plant with the hope that this strengthens the supply chains for electric vehicles (Lawson, 2022).

The UK "Critical Minerals Strategy" was published in 2022, with some updates provided in 2023. While it had been prepared earlier, key roundtables were conducted after Russia initiated the war in Ukraine. The strategy regarded critical minerals as important for energy security and military systems and noted the risks of growing demand for such materials and geopolitical uncertainties (BEIS, 2023). The key investments linked to the energy transition included the Automotive Transformation Fund (£850 million); the Energy Transformation Fund (£315 million), Energy Intensive Industries schemes, and the UK Infrastructure Bank. This is associated with some cross-coordination efforts, for instance, by the Cabinet Office having a convening role via its critical minerals' portfolio and the Natural Resources Security group operating across different ministries.

Due to not having devolved powers in security policy, it is not surprising that Scottish energy policy has paid little regard to geopolitics. For instance, the "flagship" innovation related to Scottish climate and energy policy, the Just Transition Commission, has not addressed security:

> Honestly, I have not heard word security mentioned once in any of the discussions that I have actually heard through the Just Transition Commission. Or indeed any other Scottish policy issues that I've engaged with. (Academic, 2020)

Given its devolved powers on land-use planning policy, the Scottish government has been able to consider safety and security in connection to nuclear power. From that perspective, also considering selected security aspects around the just transition could have been possible before the events of 2022.

### 8.4.2 Elements of Coordination between Energy and Security

On a more concrete level, there has been some movement toward advancing policy integration and coherence in Westminster. According to several interviewees, this was mostly visible in how security, defence, and foreign policies have integrated energy transition pursuits. The examples include climate change as a strategic agenda for foreign policy, stopping the funding of fossil fuel projects overseas, and regular high-level energy discussions in the defence administration.

This perspective is supported in the recent policy documents. For example, the 2021 "Integrated Review" mentioned measures related to energy, such as the diplomatic climate and energy network, and the need for energy transition to mitigate climate change (HM Government, 2021). Energy transition was mentioned in connection to energy security but the means to achieve this were not specified. The 2023 revised "Integrated Review" emphasized that "the transition to clean energy and net zero … is a geostrategic issue" (HM Government, 2023, p. 10). However, a real case in point, bringing the energy transition and security together, was the establishment of the new Department for Energy Security and Net Zero in early 2023, with plans to draft an "Energy Security Plan" and a "Net Zero Growth Plan."

The MoD has a climate security division. Already in 2015, the "Sustainable MOD Strategy" aimed to improve energy efficiency and reduce dependence on fossil fuels (MoD, 2015). In 2021, the MoD's "Climate Change and Sustainability Strategic Approach" recognized how climate change impacts create instability and connect to energy geopolitics:

We are already at the forefront of the new and growing green military agenda, trialing new types of vehicles, fuels standards, energy storage and much more. Done right, this will improve how we meet the defence and security challenges of the future. (MoD, 2021, p. 7)

The actions reported include helping at environmental disaster sites, using alternative fuel sources for aircrafts, and improving biodiversity in defence estates. Yet some have also expressed concern that, for instance, climate issues have been on the shoulders of individual civil servants, with limited institutional memory of the activities carried out.

Two thirds of the experts interviewed brought up issues around departmental interaction, that is, coordinating energy and security between different government ministries. Many interviewed experts observed a lack of interaction between the departments working on energy and security, defence, or foreign affairs. One of the reasons for this was believed to be the dominance of a market-oriented approach to security:

The UK does have a strong belief that the market delivers security. So, they do look at these things, but I think it's not, yeah, it's deemed as market issues and wider security issues, then energy security in terms of us having to have access to energy markets. I think within the FCDO climate change is a big issue and remains that. (Researcher, 2021)

## 8.4 Policy Coherence and Interplay

However, over time, improved policy coherence was observed:

We still operate in silos, the difference is that tops have been lopped off so we can see the different silos and we can engage with one another, so we've got much better at it. It's still not perfect. It's not seamless by any stretch of the imagination but there are lots now of, what we refer to as cross-Whitehall … cooperation between ministries. (Civil servant, 2021)

On an organizational level, a team has been placed between the ministry in charge of energy and the Foreign Office to deal with questions of international energy security. In addition, collaboration between climate and foreign affairs was perceived by some as quite strong. Yet it appears that formalized structures are still missing and, indeed, that there is organically arising collaboration. The perspective of one civil servant was that certain government divisions work on security of infrastructures and others on energy transitions, with their interconnections limited to selected meetings. Others mentioned both regular and ad hoc meetings.

The National Security Council and the Climate Adaptation and Implementation Committee were seen as formalized groups advancing coherence, and the Nuclear Skills Strategy Group as an example of more specific collaboration.

One department thought it would be a good idea to phase out fossil fuel funding overseas, that then brought in other departments all who had a view, and this came together under a committee called the Climate Adaptation and Implementation Committee, which was a gathering of, I think six or seven different ministers, chaired by the Secretary of State for Business and Energy; [the] Foreign Secretary was represented on it, the Chancellor was represented on it, so it was a fairly big, high-powered committee. (Civil servant, 2021)

There's the Nuclear Skills Strategy Group which includes someone from the MOD, someone from BEIS, someone from Rolls-Royce, but it's kind of an arm's length, so they've done this very clever thing and I think the reason they've done it is to protect from freedom of information requests, it's an arm's-length organization where every month these people get around the table and have discussions. (Researcher, 2021)

Following the events of 2022, energy, and especially energy security, have moved much higher up the political agenda. There has been consensus about cutting ties to Russian energy sources and the new "Energy Security Strategy" was published in April 2022. Despite these developments, the interviewees did not detect signs of securitizing energy policy in the UK. Generally, many remarked that the rapid changes in prime ministers slowed down policy processes.

The 2023 draft "Energy Strategy of Scotland" paid much more attention to security than previous Scottish energy policies. It emphasized developing the country's own resources, energy storage, and collaboration around the North Sea. The means outlined were, for instance, UK government-led market mechanisms and the Fuel Insecurity Fund to help struggling households. In Scotland, the highlighted security dimensions pertaining to energy and the energy transition have mainly been the socioeconomic security of its residents and nuclear safety. It seems that options to improve energy security – if not broader security – via the energy transition exist, but it is unclear if sufficient actions have been taken by the Scottish government to advance this development.

## 8.5 Niche Development, Regime Destabilization, and Positive and Negative Security

There is no niche development that substantially comes up in the security context in Scotland. As outlined, onshore wind power has become an established and significant part of the Scottish energy system. (In England and Wales, it has been "de facto" banned since 2015 due to unfavorable planning conditions.) The destabilization of the fossil energy regime is also rarely discussed in the context of positive and negative security. It has been seen to proceed as planned without major hiccups – albeit with significant implications on employment in Scotland and the need to reskill the workforce.

The offshore wind sector could perhaps be seen as a developing niche given its share is much less than that of onshore wind. Offshore wind began gaining increased support when BEIS launched the new industrial strategy at the end of 2017. A perspective arising from the expert interviews in 2020–2021 is that, with regard to the large offshore wind farms constructed on the east coast of Scotland, controversy has been raised over the substantial amount of prefabrication work conducted, for instance, in China, which has enabled lowered costs and increased scaling of renewable energy – but has increased supply dependencies on non-European actors. One reason for this supply chain dependence on Asia is due to the UK Contracts for Difference model, which looks at development costs and not the overall economic impacts of developments.[1] This, in essence, is a somewhat of a security-of-supply risk, but elsewhere it has been noted that, broadly, the development of Scottish offshore wind power benefits UK energy security (Qu et al., 2021). The energy security argument has provided needed legitimacy for the expansion of the offshore wind sector (MacKinnon et al., 2022).

Linked to the established energy regime, an ongoing concern for Scotland has been nuclear power. The Scottish government has taken a no new nuclear power policy stance, led by the SNP, from around 2006. This differs substantially from the rest of the UK's favorable view of nuclear power. In policy strategy documents from 2006 to 2010, the Scottish government saw its no nuclear policy as "a principled priority where the risk of radiation or terrorist attacks is seen larger than

---

[1] The Contracts for Difference scheme began in 2014, is operated by the National Grid, and means that low-carbon electricity generators are awarded contracts that guarantee them a "strike price" (Munro, 2018).

energy gains" (Kivimaa and Sivonen, 2021). The general perception in Scotland has been that nuclear power stations and nuclear waste transport and disposal create security risks for society. However, the opposition of nuclear power generation is also linked to the military and its weapons, in essence to the positioning of the UK nuclear deterrent submarines in Scottish waters. Previous research has argued that UK energy policy has adopted a masking strategy that has incorporated nuclear submarine construction costs into civil nuclear programs, which thereby act as a hidden subsidy for military nuclear activities (Johnstone et al., 2017). One interviewed expert stated that, even in the 1950s, the first large-scale commercial civil nuclear power plant's role was to produce plutonium for the nuclear weapons program. The Scottish government, the SNP, and civil society organizations oppose nuclear weapons, new construction of civil nuclear power, and nuclear research alike, yet these capabilities contribute substantially to the economy of some Scottish regions (Heffron and Nuttall, 2017; Ritchie, 2016).

There are about 65,000 people employed in nuclear in the UK, 30,000 of those are on the defence side. That is the biggest single share of any subsector ... The need for nuclear specialists is on the military side. You need to build submarines, that is, you need to build reactors, you need to build warheads, you have to do all these kinds of huge number of activities and, it's incredibly expensive. (Researcher, 2021)

Besides this link between civil nuclear power and nuclear weapons, new security risks related to nuclear infrastructure are emerging due to climate change. The UK civil nuclear infrastructure is argued to be

profoundly unprepared for climate impact and there is a very high probability that reactors and their associated high-level spent fuel stores will become unfit for purpose. Due to ramping climate induced sea-level rise, storm, storm surge, severe precipitation and raised river-flow, UK nuclear installations are set to flood – and much sooner than either the nuclear industry or regulators suggest. This is because risks to nuclear installations from sea-level rise driven extreme climate events will not be linear, as thresholds at which present natural and built environment coastal and inland flood defence barriers are exceeded. (Dorfman, 2021)

This perspective supports the cautious Scottish approach toward nuclear power.

What is interesting is that the exceptional events of 2022 did not change the Scottish government's opposition to nuclear power and perhaps even strengthened its no new nuclear policy, while the rest of the UK is even more strongly in favor of nuclear power. The former happened as a response to seeing the vulnerability of nuclear power infrastructure as a potential target of war. This differs quite dramatically from the perceptions of nuclear power in Finland, seen as a partial solution to reduce security risks related to energy imports (see Chapter 6). One potential explanation may be the substantial Scottish lead on the production of onshore wind power, while another links to the broader UK energy system that is able to balance nonwindy periods. Nevertheless, Scotland has much higher electricity prices than our other case countries and problems with socioeconomic security linked to energy poverty.

Fossil fuel phaseout in Scotland is coupled with renewables expansion. North Sea gas production is expected to be completely phased out by 2050. No serious national security risks are seen to relate to this development, while this could give rise to skills and employment-related risks:

I think our biggest risk, for our Scottish industry is, how quick the industry, particularly the oil and gas industry, can retool itself to service the renewable market. So, for example, Aberdeen is a centre of excellence on subsea operations, and particularly things like capping of mines and stuff, capping of wells, and pipes, undersea. How can we convert that expertise into things for, say offshore wind platforms or, subsea mining if we end up going down the subsea mining route for critical minerals, so I think the biggest risk is that the industry isn't able to evolve and retrain workers quickly enough. (Civil servant, 2021)

Finally, relatively little has been done in terms of gas security since 2022, apart from reinstating the gas storage facility.

## 8.6 Concluding Remarks

The links between hydrocarbon energy, the energy transition, and security are rather complex and manifold in Scotland and the wider UK, with relatively fragmented governance in place. While some instances of policy integration between energy and security or defence policies were found, broader policy coherence regarding security and the zero-carbon energy transition appeared lacking (Figure 8.4). This links to the somewhat conflicting intertwining logics of energy and security regimes, which simultaneously support the old energy system while also aiming to change it. In addition, before 2022, many of the efforts that relate to coordination across energy and security regimes were focused on external or global energy questions – such as safeguarding fossil fuel trade routes or advancing energy diplomacy via renewables – instead of domestic energy policy. The events of 2022 seem to have raised questions of security and energy transition links in political and policy agendas pertaining to domestic energy production and use. Figure 8.4 summarizes the key aspects of the energy–security nexus in Scotland and the wider UK and their effects on the energy transition.

Attention paid to broader security questions in energy policy was low, especially between 2016 and 2021, or was sometimes masked from the wider public, as in the case of civic and military nuclear power. Generally, the large influence of the private sector on energy policy and the dominance of economic and market-based values – essentially an active and deeply ingrained depoliticization of energy – have

## 8.6 Concluding Remarks

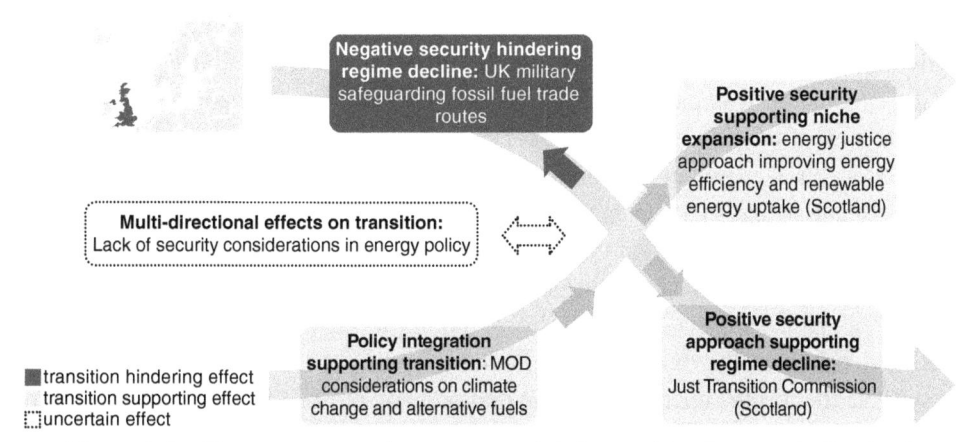

Figure 8.4 Key energy security aspects and their transition impacts in Scotland and the wider UK, 2006–2023.
Source: Kivimaa, Finnish Environment Institute, 2023.

led to relatively little attention being paid to the connection between security and energy transitions prior to 2022. Some argue that insufficient attention was paid to the expansion of new niches, such as energy storage and smart grids, as means to improve energy security, and that there has been an overreliance on the privately owned National Grid in energy security. However, there is rising interest in these issues, in particular critical materials security.

At the level of objectives, in key policy documents some interconnections were made between energy and security in the early 2000s. These reduced in emphasis after the Conservative government came in power in 2015, alongside dilution of earlier ambitious low-carbon energy policies. In addition, a divergence has existed between a hydrocarbon-based security approach to safeguard international fossil fuel routes (which has been rather coherent) and the advancement of climate security via the energy transition. The former represents a kind of negative security-based thinking where physical attacks to fossil fuel flows were prevented (see Hoogensen Gjørv, 2012; Hoogensen Gjørv and Bilgic, 2022), while the latter could benefit broader positive security thinking based on enabling communities via the energy transition to be prepared for climate change impacts. The latter could also connect better to the Scottish just transitions agenda. The economic focus of energy policy has had a negative influence on cross-departmental coordination of energy and broader security issues. Such collaboration is seen to be increasing but, at least before 2022, it has tended to be more organic and ad hoc than formalized.

Geopolitical concerns over Russia were not particularly important for Scottish or wider UK energy policy pre-2022, as the island is geographically remote from

Russia. Even some of the energy experts understood the indirect interconnections only after the 2022 events. Yet the energy transition, and especially wind power niche expansion, was partly legitimized based on the need to reduce the indirect dependency of the UK on Russian energy flows. Nevertheless, prior to 2022, the discussion on security in UK energy policy was mostly limited to the kind of energy security that markets can deliver or "real-time" security-of-energy supply. While the political rhetoric of energy security became more visible after the 2022 events, little concrete changes in policy instruments were evident in early 2023. However, critical materials availability – linking to growing expectations of geopolitical risks and the need for new learning around the energy transition – is a new issue that brings energy transition and security closer together than before. In the UK, this has mainly been tackled by investments in innovation and domestic lithium production – that is, new niche development.

Scotland has had a different worldview on security in relation to energy transition than the broader UK. While in-depth explorations of these links have been lacking, more attention has been paid to the environmental and health security effects of energy policy choices and just transitions – the latter linking broadly to the conceptualization of positive security. These have been evident, for instance, in the rather long opposition of nuclear power, opposition to fracking, and more consideration of energy poverty and the just transition than in the rest of the UK.

# Part III

Conclusions

# 9

# Insights into Zero-Carbon Energy, Sustainability Transitions, and Security

Chapters 5–8 delved deep into four case countries: Estonia, Finland, Norway, and Scotland/the UK. This chapter looks at the empirical findings of this book in a comparative light. It does so by using conceptualizations introduced in Chapter 2, such as negative and positive security, securitization, and politicization, as well as by focusing on the analytical dimensions of interest in Chapter 4: coherence and integration between energy (transition) policies and security and defence policies, security as part of the landscape for energy transitions, and security in niche expansion and regime decline processes.

The chapter also aims to answer the questions presented in Chapter 1: What are the security implications of energy transitions? What elements of positive and negative security can be found? How should energy security and security of supply be redefined in the context of the energy transition? Is there a hidden side to policymaking with regard to the energy–security nexus? It first discusses the interplay between energy, security, and defence policies, followed by securitization and politicization. Subsequently, focus is placed on the security implications of energy transitions and negative and positive security. The chapter ends by summarizing the key technological, actor-based, and institutional aspects of the country cases, looking at Russia as a landscape pressure, and then providing final conclusions.

## 9.1 Interplay between Energy, Security, and Defence Policies

As explained in Chapter 4, the research conducted for this book approached policy coherence and integration in different ways. On the one hand, it looked at processes and measures that aimed to integrate security into energy policy – and vice versa. On the other hand, it examined synergies and conflicts between energy (transition) policy and defence and security policy. Before 2022, one can observe low or moderate levels of policy integration in the two domains under scrutiny, insufficient administrative interaction, and conflicts between the objectives and means of advancing the zero-carbon energy transition and the objectives and means of national security and defence policies.

Table 9.1 summarizes the results of the country cases. The level of policy integration has varied across countries but also between policy domains. Estonia has had the highest level of integration between the objectives of energy and security policies but has still suffered from informal administrative interaction and conflicts between the implementation of energy and security policies – most visible in the problems related to the phaseout of oil shale and the expansion of wind power on security grounds. In Finland, policy integration has ranged from low to moderate. Both policy domains' strategies have mentioned supporting integration and included some measures, such as the Power Pool (see details in Chapter 6) or assessments of the effects of wind power on defence radars. However, administrative coordination has been fragmented and tensions have existed, exemplified by the difficulties of expanding wind power to certain parts of Finland, the justification of peat energy for reasons of energy security, and the avoidance of discussing geopolitical risks pertaining to Russian energy collaboration before 2022. In Norway, policy integration and coherence have been on a low level because security was largely a nonissue in relation to energy policy before 2022, while economic security provided by oil and gas exports gave continued support of this direction. In Scotland/the UK, there has been a relatively high integration of energy and climate change issues into security and defence policy strategy documents, but integration of security into energy policy has been modest and the coordination of energy and security policies fragmented. In general, the perceptions of risks in energy policy have been more focused on economic–political aspects than technical and physical risks from military or terrorist attacks.

When one looks at how policy integration in this nexus has changed since 2022, there is evidence of learning-based integration in the case countries. First, many expert interviewees reported gradual improvement of the interaction between the administrative sectors in charge of energy, security and defence. And, for instance, the role of the North Atlantic Treaty Organization (NATO) in building awareness about climate change within defence contexts was also noted more broadly.

Despite improvements, one problem for policy coherence is due to differing values and worldviews between domains – often unresolved at the political level. For example, the prioritization of different policy objectives varies between sectors: Defence policy actors emphasize operational capability of defence as the most important factor, whereas energy policy actors highlight the secure provision and price of energy and its carbon dioxide emissions. As an illustration, the expansion of wind power, to complement other energy sources, improves energy availability and reduces prices in many places. However, it hinders the operational capability of the defence sector in cases when wind turbines prevent accurate air surveillance imaging. Nevertheless, deepening learning and networking (see Ghosh et al.,

Table 9.1 Summary of policy coherence and integration before and after 2022

|  | Estonia | Finland | Norway | Scotland/the UK |
|---|---|---|---|---|
| Pre-2022 | Moderate-to-high integration with help of market collaboration with Europe, diversity of supply, oil shale, desynchronization from Russian grid. But insufficient formal collaboration between ministries and conflict between wind power expansion and defence radars as well as oil shale phaseout aiming for decarbonization and (perceived) security. Synergies via the desynchronization project. | Low-to-moderate integration, lack of concrete holistic policy measures, fragmented collaboration, and conflict between wind power expansion and defence radars. Yet, coordinating elements with potential, for example, national security-of-supply organization, power pools composed of public and private actors, and the comprehensive security model. Synergies more generally in terms of phasing out fossil fuels and energy security. | Low integration, nonidentified links between energy and security. Some emergency preparedness and cybersecurity measures (coordinated by Norwegian Water Resources and Energy Directorate (NVE)) that involve energy and security actors, but not in a major role. Conflict between economic security and geopolitical influence of Norway with zero-carbon transition. Synergies not recognized in expert interviews. | High-to-moderate integration of energy and climate change into defence and security policy. Low-to-moderate integration of security in energy policy, mainly limited to energy security. Conflict between zero-carbon transition and the military securing the fossil fuel trade. Synergy in terms of expanding wind power in Scotland and (perceived) improved energy security. |
| Post-2022 | Improved collaboration regarding wind power expansion and defence; conflict of maintaining oil shale (receiving increased political consensus due to energy security) versus energy transition; acquiring liquefied natural gas (LNG) terminal with Finland as an example of exceptional measures. | Moving from diverse views to consensus about Russian risk and increased attention to security in energy policy. Extraordinary policy measures, for example, LNG terminal with Estonia, discontinuing Russian energy imports, emergency stockpile of peat, reinvestigating opportunities to increase wind power in eastern Finland. Generally, increased recognition of energy security provided by the energy transition. | Increased focus and connections between energy and security, for instance, pertaining to military actors securing critical energy infrastructure and the vulnerability of infrastructure to intentional attacks and climate change effects. Conflict in terms of Norway's importance for European energy security and decarbonization. | Some examples of increasing interaction between energy and security (e.g., "Critical Minerals Strategy") and more focus on energy security from a geopolitical perspective, but overall coherence still lacking. |

2021) across defence and energy sectors have been paramount in partially resolving the conflict between wind power expansion and defence radar operation. It is important to note that achieving perfect coherence is often impossible. Improving coherence to advance chosen trajectories does, however, mean that some policy objectives or measures may need to be abandoned. For instance, stockpiling fossil fuels becomes an impossible energy security measure when energy transitions are advanced. Or securing the operational capability of defence forces may mean that electrifying a country's military fleet cannot be an objective for defence policy.

Unlearning established practices (see Van Oers et al., 2023) in the energy administration seems vital in order to take a new updated approach to the energy–security nexus. This also means a disruption of existing skills of both civil servants and energy businesses and a search to find areas in which existing skills can be repurposed (Kivimaa and Sivonen, 2023). Expectation dynamics played a relatively small role in the energy–security nexus before 2022. For instance, there was relatively little discussion on critical materials security in relation to expanding renewable energy at that time.

One explanation for the incoherence between energy (transition) policy and security and defence policies is the low political importance assigned to such coherence before 2022 in all the case countries, apart from Estonia. This seems to be the case in many other Western and Northern European countries too (Kuzemko et al., 2022). This incoherence has partly been affected by the depoliticization of energy (or at least certain energy sources) and, in cases, even by its desecuritization (see Section 9.2). Despite some modes of coordination, policy outputs and outcomes have often been incoherent. As a follow-up to the 2022 events, the interconnections between security and energy have become some of the key topics in the media and policymaking alike. As a result, policy integration and coherence are likely to improve but require an explicit recognition of the connections – both synergies and conflicts – in the implementation of policies in both domains. The increasing debate on climate security generally (Busby, 2022) and its growing focus in security policy (Farham et al., 2023) create opportunities to find improved alignment between the two domains.

In conclusion, improvements in policy coherence are needed on many levels to accelerate energy transition and do it with security questions in mind. First, explicit identification of synergies and conflicts between the energy transition and security and defence policies is required so that attempts can be made to resolve potential conflicts and improve synergies. Some issues of increasing importance deal with long-term trajectories for fossil fuels and the material dependencies related to the expansion of renewable energy. Second, administrative interaction between the policy domains is essential, with formalization of processes that improve the transparency of policymaking outside the energy elite to the broader society. Third, improved focus is needed on learning-based processes to support policy integration in a rapidly changing world.

## 9.2 Securitization and Politicization of Energy Transitions

The concepts of securitization and politicization and their nuances were explained in Chapter 2. In one interpretation, energy was mostly depoliticized and desecuritized prior to 2022 in the case countries – apart from some specific questions of politicization around peat in Finland. Evidence of securitizing moves and audience acceptance was not found in the interview data, following the classical definition of securitization by the Copenhagen School of Security Studies. This can be explained by the principal market logic of energy policy, as well as the technocratic perspective that often dominates in the energy sector.

In another interpretation, if one thinks about securitization in a lighter manner, for instance, as described by Johnstone et al. (2017) as altering policy goals in terms of military-oriented national security, one can see signs of securitization in Estonia in terms of the strong pursuit to disconnect from Russian energy flows and the influential role of the transmission network owner Elering. Estonia can, however, be regarded as an outlier among the case countries, because security has been a standard part of its energy policy for many years, as in many other Eastern European post-Soviet countries. In the UK, connections between military and civic nuclear power, as reported by Johnstone and colleagues, hint toward securitization, but in many respects energy policy in the UK has been associated with both desecuritization and depoliticization. In Finland, attempts have been made to keep geopolitical considerations of energy nonpoliticized and energy policy desecuritized. In Norway too, energy questions appeared desecuritized before 2022.

For a third interpretation, Heinrich and Szulecki (2018) have proposed three dimensions of securitization in the energy context: exceptional measures, strengthening the executive powers of selected agencies, and isolating selected decisions and potentially important information from public access. None were particularly evident before 2022. Regarding the latter, most interviews did not reveal a consciously hidden side to policymaking in the energy–security nexus – although those interviewees outside the energy elite would not know about the hidden side. There was one reference to the previously hidden connections between civil and military nuclear power in the UK (see Johnstone and Stirling, 2020). Some of the issues identified in this study appear to have been "public secrets," such as the geopolitical risks Russia posed to the Finnish energy sector. On occasion, they have been discussed in the media by selected experts but omitted or ignored as unrealistic by others. Generally, the lack of discussion on the energy–security nexus was very observable before 2022. The informality of the exchanges between energy and security administrations have reduced transparency and could in principle be seen to be contributing to securitizing energy policy, because the few discussions and decision-making that have taken place in the nexus have been hidden from the public eye. This setting changed since 2022, the events of which politicized energy.

In 2022, the invasion of Ukraine by Russia and the ensuing energy crisis made energy transitions more strongly politicized, especially concerning energy prices and availability. These events did not appear to lead to securitization as defined by the Copenhagen School. There has, of course, been increasing collaboration across the policy domains of energy and security. The policy measures taken do not appear exceptional to the extent defined in security studies, while again, if interpreted in a somewhat lighter manner, they do amount to extraordinary measures that break with normal political practices (see Heinrich and Szulecki, 2018). Yet decisions have been made in ministries that have been more open for public scrutiny than before, as the crisis increased the interest of the public on energy matters, making energy politicized. Hence, the post-2022 situation in the case countries does not match with all three elements of securitization proposed by Heinrich and Szulecki (2018).

Politicization of energy has a beneficial dimension. According to security studies, environmental issues should be politicized, if they are not securitized, to make sure they will be addressed (Floyd, 2019; Trombetta, 2009). This will create openness and transparency with regard to decision-making. Politicization is particularly important in the context of energy transitions and security, because the case studies showed how security can be used both as an argument for and against sustainability transitions. However, politicization also creates risks that relate to, for instance, important decisions being made within short timeframes and with the motive of appearing popular to the electorate. By politicizing decision-making, but making sure decisions are based on the latest scientific knowledge, security implications of the transitions can be best assessed.

## 9.3 Security Implications of Energy Transitions

The security implications of energy transitions, based on the views of interviewed experts from the case countries, were analyzed in detail in a scientific article (see Kivimaa and Sivonen, 2023). I summarize here some of the key elements and connect them to the processes of niche development and regime decline as well as the conceptualizations of negative and positive security described in Chapters 2 and 4. To recap, negative security refers to the traditional understanding of security against the appearance of threats, whereas positive security emphasizes people's feelings of being free from insecurity, emancipation and empowering individuals and communities (Booth, 2007; Hoogensen Gjørv, 2012; Hoogensen Gjørv and Bilgic, 2022).

In the case countries, the negative security approach toward energy security has traditionally been rather prevalent; that is, in terms of maximizing the production of domestic energy (typically fossil fuels) and stockpiling fuels in case of crises. The former was especially visible in Estonia, whereas Norway has sufficient hydropower for domestic energy consumption. The latter has been typical of Finland and is growing in importance in Estonia. For Norway, stockpiling has not been a concern, whereas the UK only reinstated its gas reserve recently. In addition, energy security has involved preparedness for military and other types of physical and cyber risks, although the events of 2022 and 2023 illustrated that the risks for critical infrastructure had been underestimated. In the UK, the involvement of defence sector actors to safeguard global fossil fuel trade routes is another illustration of the traditional negative security approach in this nexus. Nevertheless, attention toward negative security has been limited because market logic largely prevailed over security-oriented thinking.

The term positive security was not used explicitly in policy documents or by the experts. Instead, the associations with positive security were explored via the assumptions, practices, and actors in the case countries. For instance, renewable energy is often associated with positive security. Especially when decentralized, it can enable local communities and improve local energy resilience. In this way, it creates freedom from insecurity (see Booth, 2007). Examples of positive security in the case countries included Scotland's Just Transition Commission and policies to reduce energy poverty. In addition, the EU's Just Transition Mechanism has been applied, for instance, in the regions of oil shale production in Estonia and peat production in Finland to support the energy transition and alleviate its negative consequences. The Norwegian Sovereign Wealth Fund is an example of positive security creation associated with fossil fuels, and hence it is feared that the phase-out of fossil fuel production in Norway will reduce societal security. Nevertheless, broadly, countries orienting their energy policies toward just energy transitions, citizen participation, and energy democracy are more likely to align with positive security – with added potential to combat internal security risks arising from fossil fuel phaseout and populist politics.

The areas in which security was seen by the interviewed experts to be affected by the energy transitions in different ways included: energy security, electricity system operability, geopolitics, defence, cybersecurity, and internal stability. The research conducted did not analyze the magnitude of these risks, which have been noted to differ (see Winzer, 2012).

In terms of energy security, there were widely shared expectations in the four case countries that the expansion of renewable energy niches will improve self-sufficiency, where new technical solutions alongside local energy communities (with potential for positive security) will continue to improve energy security. However, there are also technical and institutional risks involved, including the reliability of renewable energy sources, the availability and price of critical materials, and the functioning of new kind of network dependencies. The analyses presented in this book showed that explicit assessments of such benefits and risks were largely not conducted in the case countries before 2022, whereas they were of interest to the EU and international organizations such as the International Energy Agency (IEA). The decline of the fossil fuel regime is also an energy security concern, especially for those countries with domestic hydrocarbon resources. For instance, the phaseout of oil shale has been difficult in Estonia, because it reduces the country's energy independence before renewable energy becomes more widely adopted. In contrast, the UK coal phaseout has been such a long-term process that it is no longer seen to substantially impact energy security. The security implications of emerging energy niches (e.g., green hydrogen) were largely unexplored.

Broader energy security effects are linked with electricity system operability, which becomes more important with the advancing electrification of society. The expectations regarding this were not consistent, ranging from the system becoming too risky to containing mostly solvable challenges. It was emphasized that the transition will require new learning and increases other actors' dependence on large universities and global companies regarding technical solutions. Institutionally, there was an expectation of increased cross-border reliance on neighboring countries via interconnected electricity systems. The existing electricity interconnections between the case countries, forming new and expanding grid communities, address part of this risk. Yet there are many questions around electricity storage and variable pricing related to system operability. In this context, electricity interconnections can be connected to broader questions of geopolitical alliances, that is, with which countries does one choose to build such systems. The interconnections are influenced by geography, but the Estonian desynchronization project shows that foreign policy decisions too can be made regarding such issues. The interconnections (or their lack) also indicate the willingness of states to collaborate (or not) with other states in the advancement of the energy transition, while interconnections also mean new electricity export opportunities.

The geopolitical implications of energy transitions are likely to be manifold, and have already been rather extensively covered in Chapter 3. The large hydrocarbon-based conflicts are expected to reduce, while new types of conflicts around critical materials are emerging relating to the relations of the EU with the rest of the world, especially China. Connected to resourcing such materials from the Global South, there are many examples of negative impacts on the environment, health, and human security – that is, declining positive security more locally (e.g., Sovacool, 2019). In turn, sourcing materials from the Global North depends on the setting. In some cases, positive security can occur via improvement of local communities, while there are also many risks, for instance, related to the "resource colonialism" of the Sámi lands (e.g., Sörlin et al., 2022) or local environmental destruction. The case countries of this book are likely to benefit geopolitically from renewable energy, due to their high technological competences (Kivimaa and Sivonen, 2023).

With respect to the broader dimensions of security, starting with defence there were expectations that wind power is broadly beneficial to national security and micro-grids offer military operations more security. Such issues have also been acknowledged elsewhere to accelerate energy independence in connection to military combat (Farham et al., 2023). The experts highlighted emerging opportunities via military research and development (R&D). In turn, technological and institutional learning have already enabled some of the conflicts between defence radars and wind power to be resolved, as evidenced in Estonia and Finland. The security implications of the transition in the defence sector are mainly connected to the negative, that is, hard security, perspective. Experiences from the case countries showed that improved dialogue and interactions are needed between energy and defence sector actors on this topic. NATO has been oriented toward this theme for some time and can advance discussions across countries (Farham et al., 2023).

The electrification of the energy system and expansion of renewable energy are connected to more digitalized and complex systems. Digitalized systems are expected to increase vulnerability to cyberattacks, for example, such that described in the case of Norsk Hydro (see details in Chapter 7). Whereas physical power plants are typically not connected to the Internet, their office systems may be subject to attacks. The interviewed experts also emphasized the need for civil servants and companies to learn more about cybersecurity and collaborate more broadly.

Finally, energy transitions can affect countries' internal stability by creating tensions around fossil fuel phaseout and fast-advancing niche expansion. Experts in Estonia, Norway, and Scotland/the UK referred to a risk of tensions and unrest created by livelihoods threatened by climate change goals. In Estonia, an added problem is that oil shale production is located in a region of economic hardship and high unemployment. Therefore, attention needs to be paid to compensation, and to retraining and repurposing fossil fuel industry skills and assets; this is where the just transition initiatives can help. Another area of potential tension is inequality between people's ability to benefit from the technologies associated with transitions, such as solar panels, heat pumps, or electric vehicles. These technologies may be unavailable to those on low incomes or living in rented accommodation. This links to energy poverty, a key policy area in Scotland. Tensions around the energy transition can further escalate, because many political far-right (or sometimes far-left) parties are working to resist decarbonization efforts (Vihma et al., 2021) and use social disruptions for political gains. Instead, increased measures oriented toward social justice and just transitions are needed and can alleviate some of the tensions.

With respect to the internal and external dimensions of energy security listed in Figure 3.1 (Chapter 3), it is easy to see that internal dimensions have had a long-term presence in all countries' energy policies. These include access to energy, affordability, diversity of sources, degree of self-sufficiency, nondependence on a specific geographical region, and resilience to shocks. The external dimensions that relate to broader security have perhaps been less considered, apart from impacts on climate change. For instance, impacts on welfare and energy justice have been considered for some time in Scottish energy policy, but in the other case countries only emerged because of the 2022 energy crisis. Risk of military and terrorist attacks have been acknowledged for many years in the case countries' policy strategies, but the risk has only fully realized since 2022. Security and supply of materials and components necessary for the energy transition and the effects of renewable energy deployment on land use have mostly only become considerations in the last few years. In conclusion, policy coordination needs to improve so that the external dimensions of energy security and the range of security implications described in this book become integral parts of energy policymaking.

## 9.4 Technological, Actor-Based, and Institutional Aspects

In this section, I summarize and compare the case country findings related to the categorization of technologies, actors, and institutions (see Chapter 3). Regarding *technological* aspects in this nexus, the largest commonality between the case countries are the security implications of the expanding wind power niche, albeit in differing ways. In Estonia and Finland, the key issue has been the effects of wind turbines on the operation of the defence sector's air surveillance radars due to the closeness of the Russian border. In Scotland, wind power is seen to improve energy security and replace fossil fuels, but the prefabrication work of the turbines in China has caused some concern. In Norway, the expansion of wind power is countered by an antiwind power movement that has created societal tensions. Another emerging commonality is securing critical energy infrastructure against military and terrorist attacks – an increasing concern since 2022 and 2023, which witnessed the explosions targeting the Nord Stream gas pipelines and the damage caused to the Baltic Connector gas pipeline between Estonia and Finland. Such events also exposed the vulnerability of fossil fuel infrastructure to attacks and indicated that renewable energy and local energy solutions can improve energy security. Interestingly, the findings also showed that, during 2020–2021, the governments of the case countries had paid little attention to the security of supply of the critical minerals and metals required by the expansion of renewable energy and energy storage solutions – something that has definitely changed since.

What has become clear is that many security issues connected to the zero-carbon energy transition do not seem important or are not widely discussed until technological niches begin to substantially expand, as shown by the case of wind power, for example. In particular, when niches move from modest fit-and-conform empowerment to much more disruptive stretch-and-transform empowerment that changes the sociotechnical energy system (see Smith and Raven, 2012) the potential implications become much more visible in the security regime and can be possible places of tension and contestation. Therefore, a more future-oriented approach toward analyzing the possible security implications of expanding sustainability niches would be useful in policymaking. Perhaps the current decade of crises has already included security among the expectation formation and learning processes for new sustainability niche development, but it is important to make sure a more long-term anticipatory perspective is truly adopted.

In all the case countries, the destabilization of the fossil fuel-based regime is also somehow affected either by security concerns or, at least, by security rhetoric. The countries, however, revealed divergent issues around fossil fuels and security. In Norway, the export of oil and gas has not only strengthened the country's economic security – bringing positive security to the whole society via the Sovereign Wealth Fund operated by fossil fuel income – but also made the country geopolitically more influential than its size would normally allow. Despite the economic importance of fossil fuels in Norway, the offshore wind sector also provides opportunities to repurpose skills from the hydrocarbon sector for a more managed regime decline. In Estonia, oil shale has provided energy independence from Russia, but its phaseout has also led to concerns over Russia's reaction, because the oil shale production region has a large Russian-speaking population and is close to the country border. Here, the EU Just Transition Mechanism has been used to create new industry and potential positive security for Ida-Viru County, for example, by supporting a new magnet factory producing components for the energy transition. In the UK, fossil fuels have more generally been tied to the operation of the military in safeguarding international supply routes, although the UK too was affected rather substantially by the gas crisis following Russia's full-scale invasion of Ukraine in 2022. Particularly in Scotland, the Just Transition Commission has sought ways to improve positive security, for instance, by reskilling fossil fuel workers. Questions of phasing-out production have, however, been raised at a lesser scale than in Estonia and Norway, perhaps due to the decades-long UK coal phaseout (see Turnheim and Geels, 2012). Nonetheless, the oil and gas sector in Scotland and its future were under lively discussion at the time of writing, with decisions pending. In Finland, domestically produced peat (while not a fossil fuel it produces greenhouse gas emissions equivalent to fossil fuels) has been framed in terms of energy security. Here too, the EU Just Transition Mechanism is used to seek opportunities to repurpose the skills and assets of the peat industry and its workers. Regarding exported fossil fuels, there seems to be consensus about the feasibility of phaseout.

The technological characteristics of the sociotechnical energy regimes have coevolved with security regimes over time. This means that creating more synergies between energy transition policies and security and defence policies is needed. The Estonian country case illustrated an interesting example of coevolution by solving the conflict between the operation of the air surveillance radars and wind power by constructing more efficient radars – although this was a result of a rather long process that also involved tension and conflict. In Finland, the public rhetoric since 2022 has largely framed a synergistic relationship between wind power and national security, and has aimed to speed up wind power permitting, indicating perhaps a cultural–institutional coevolution between security and energy regimes (cf. Grin et al., 2010). With respect to Norway and Scotland, it was harder to observe coevolution of energy and security regimes before 2022. Some examples in the UK may be the gradual changing of the defence regime to better account for climate change and the specific ties between civic and military nuclear power (Dorfman, 2017; Johnstone and Stirling, 2020). However, many political efforts since 2022 have been oriented toward better fitting together the energy and security regimes. In many European countries, the landscape shock of 2022 when Russia invaded Ukraine has led to a realignment pathway (plans to develop green hydrogen and small modular nuclear reactors) and a technological substitution pathway (expansion of wind and solar power) (cf. Geels and Schot, 2007).

The *actor* dimension is connected to the power to advance or hinder things (i.e., "power to"), dependencies between actors (i.e., "power over"), and the power of coalitions of actors ("power with") (Avelino, 2021). This study of the energy–security nexus shows the interdependencies of actors, where sometimes security actors have power over energy actors when the question is vital to national defence – such as the effect of wind power turbines on air surveillance. However, most of the time the energy sector actors have had the power to ensure economic reasoning prevails. The Finnish case, interestingly, also revealed the power that politicians possess over civil servants, by hindering discussion about the geopolitical risks of energy imports from Russia prior to 2022. It is also important to note that different actors had differing perceptions of the energy–security nexus and the power of actors. One potential reason for this was the division between those that belong to the inside "energy elite" (see Ruostetsaari, 2010, 2017) and those outside it. Those on the outside are not, for instance, aware of any informal interactions that take place.

Although the case countries have long traditions of climate and energy policymaking across different ministries, often such processes have not involved actors in defence or foreign affairs, except in Estonia. Generally, the countries studied in this book were mostly relying on informal interaction between the ministries responsible for energy and for security and defence. This was argued to work well in small country contexts. It has also meant that the role of security actors has remained rather implicit in energy transitions. Sometimes, such actors – for example ministries of defence – have slowed down energy transitions due to valid concerns about the impact of wind turbines on air surveillance radars. At other times, security actors have been excluded from energy policymaking, with argumentation related to the market orientation of energy policy or the avoidance of "securitizing" energy policy (Kivimaa, 2022). The country analyses also showed that it was important to include private sector actors in discussions at the nexus of energy transitions and security. For instance, energy business actors are likely to have more up-to-date and accurate information about the range of security issues that energy transitions involve and what the solutions could be – but security-sensitive government information cannot be disclosed to them unless they are included in such discussions. Yet, some business actors may also have (too) strong roles in energy policymaking, as illustrated, for instance, by Eesti Energia (see Chapter 5 for details). This links to the role of the state in the energy sector, discussed later in this sub-section.

Actors connect to *institutions* and the arrangements constructed to govern the interplay between energy and security. The country cases portrayed examples of institutions at this interface, for instance, security-of-supply organizations coordinating stockpiles of fuels and emergency protocols in case of electricity system disruptions. These institutions, however, seemed to be rather narrow in focus, typically excluding broader military security or geopolitical concerns. Some case countries did not even have some of these institutional structures in place. What the energy transition entails is rethinking security of supply within the context of the increasingly electrifying energy system with a larger share of intermittent renewable energy, which makes stockpiling difficult. Therefore, new institutional structures are needed around technologies and business models that consider what energy security means in the context of a new decarbonized energy regime. With regard to the electricity sector, it is vital that such institutions also reach across country borders to enable positive collaboration in the supply and transmission of electricity.

As noted, in the case countries informal institutional arrangements were more common than formalized arrangements across the energy–security nexus. This may be important in the sense that institutions exert influence, guiding behavior and perceptions. Societal actors may have been less aware of/prepared for security risks facing the energy infrastructure – affecting energy availability and prices to end consumers – when such questions were not part of formal institutional arrangements and, hence, not openly discussed. This could be seen in the reactions to the 2022 energy crisis in Europe. The findings also showed that informal rules at the intersection of energy and security regimes have at times hindered energy transitions.

Examples include the dissonance between how decarbonization and energy markets relate to security (in Finland and Norway) and the idea that markets best deliver energy security (in the UK). In effect, informal institutional structures have carried the responsibility of coherence – or lack of coherence – between energy and security policies. Whereas informality may often work well, it also means that there is a lack of accountability and transparency on behalf of the decision-makers and the public administration. Given the seriousness of both climate change and broader security concerns for societies, formal institutionalization of this interface in support of resilient zero-carbon energy transitions is required.

The institutional aspect of the energy–security nexus also connects to the role of the state in the energy regime and the energy transition (see Johnstone and Newell, 2018). Within the four country cases, the role of the state varied in the countries' energy–security nexuses. In Norway, it was the strongest, due to large government ownership of energy production (hydropower and fossil fuels) and almost exclusive ownership of electricity transmission and production. This is not necessarily most conducive to zero-carbon transitions, because the Norwegian state also has an interest in maintaining fossil fuel production, which provides economic security for the country. It also means that energy transitions are mostly advanced by large regime actors, such as Equinor, the largest fossil fuel producer in Norway. Another example comes from Estonia, where Eesti Energia functioned as state-owned monopoly until 2014, producing electricity from oil shale and being very influential on Estonian energy policy (albeit since then it has begun to orient toward the energy transition). In Finland, energy production and transmission have not been owned by the state to the extent they are in Norway. The Finnish state has, however, played an active role in advancing the energy transition, although changes in the government and voices of antitransition could change this. This means, among other things, that the actions of private sector actors can also be beneficial for the energy–security interface. Private ownership of energy production can be more conducive than state ownership to advancing the transition. Moreover, in Finland, the expansion of renewable energy and electrification has been associated with improved security, due in part to the lack of domestic fossil fuel production. On the other hand, high private sector dominance can also be unconducive to zero-carbon developments. The Scotland/UK case shows that high private ownership of, for instance, transmission network capacity may slow the prerequisites for energy transitions, while state actors, such as the Office of Gas and Electricity Markets (Ofgem), have also been rather reluctant to advance the energy transition. Therefore, it seems that balanced roles of public and private as well as regime and niche actors work best for both the advancement of zero-carbon transitions and the security of the sociotechnical energy system. Table 9.2 summarizes the key aspects in the case countries' energy–security nexuses.

Table 9.2 *Summary of country findings regarding technologies, actors, and institutions*

|  | Estonia | Finland | Norway | Scotland/the UK |
|---|---|---|---|---|
| Technologies | Wind power niche expansion hindered by their effects on defence radars.<br>Destabilization of oil shale regime slow due to risk of Russian reactions, economic security in the region, and security of supply.<br>Desynchronization from Russian energy network a long-term energy security process.<br>Emerging attention on critical materials. | Wind power niche expansion hindered by their effects on defence radars.<br>Russian energy company as investor in a new nuclear power development (halted in 2022).<br>Peat (part of fossil energy regime) framed as important for energy security, while actual contribution low.<br>Emerging attention into critical materials and security of critical infrastructure since 2022–2023. | Economic security and geopolitical influence tied to fossil fuel exports.<br>Most attention in security terms to the safety and risks related to hydropower dams.<br>Some tensions around expansion of the wind power niche – a question of antiwind power movement.<br>Security of critical infrastructure an emerging concern from 2022. | Nuclear power opposed due to security of infrastructure reasons (by Scotland).<br>Critical minerals (required by renewable energy) a concern for some time in the UK, increasing focus from 2022.<br>Wind power seen to improve energy security, but Chinese prefabrication work has raised some concern.<br>Military safeguarding fossil fuel supply routes. |
| Actors | From dispute to cooperation between ministries in charge of energy and of defence.<br>The network operator Elering described as "quasi-security police."<br>Divergence in how expert interviewees perceived interaction between energy and security actors.<br>Consensus between political parties on importance of oil shale post-2022. | Defence and foreign affairs ministries not part of coordinated climate and energy policymaking.<br>Energy elite actors have emphasized competitiveness over geopolitics.<br>The Power Pool was found to be the most efficient actor network coordinating part of the energy–security nexus.<br>Long tradition of public–private collaborations. | Interconnections between energy and security actors were largely not seen as important prior to 2022.<br>Large state ownership of hydropower, oil, and gas installations creates tight state–business relations.<br>NVE coordinates power supply preparedness and cybersecurity.<br>Antiwind power movement actors questioning "energy sovereignty." | Large role of and reliance on private sector actors (and markets) in energy markets and security.<br>Defence sector actors have had traditional roles in safeguarding fossil fuel trade routes but are also considering impacts of climate change.<br>Selected (rather fragmented) fora, where energy and security actors meet. |

Table 9.2 (cont.)

| | Estonia | Finland | Norway | Scotland/the UK |
|---|---|---|---|---|
| Institutions | An informal approach to interministry coordination has dominated, leading to potentially ad hoc problem-solving. Lack of formal institutions at the energy–security nexus (esp. before 2022). Military spending 2.1 percent of GDP (2022). | A collaborative governance model has created networks supporting energy–security policy coherence, but no one has been assigned responsibility for the whole energy–security nexus. Military spending 1.7 percent of GDP (2022). | Heavily institutionalized role of oil and gas income; related four ministries in the energy–security nexus not sufficiently coordinated. Insufficient institutional structures for the energy–security nexus. Military spending 1.6 percent of GDP (2022). | Lack of formal energy and security policy mandate prevents Scotland's coordination of energy–security nexus, but partly divergent institutions from the UK via spatial planning powers. Military spending 2.2 percent of GDP (2022). |

## 9.5 Further Insights for Sustainability Transition Studies

The invasion of Russia in Ukraine had a large external impact on energy policy in the EU and its member states in 2022 (Kuzemko et al. 2022). The resulting actions of the European Commission led to the halting most of the coal, oil, gas, and electricity flows from Russia to Europe. This reduced the availability of energy in European countries and resulted in skyrocketing prices of electricity, heat, and petrol. The event can be described as a security-related "landscape shock" for the European energy regimes.

The findings from the case countries show that before 2022, energy and security experts had differing perceptions of Russia as a landscape pressure on the energy sector. These ranged from perceiving a substantial risk to remarks about low risk and good energy collaboration. While the views of the experts were mixed, broadly most energy experts in Finland, Norway, and the UK had relatively few concerns and viewed the energy collaboration positively. Estonian energy experts – and security experts in all case countries – tended to have a more cautious perspective.

The annexation of Crimea in Ukraine by Russia in 2014 caused a small landscape shock, where the Russia risk was brought to the fore more strongly in the case countries. However, this had relatively little impact. It did not result in markedly improved coherence between energy and defence policies. In turn, the substantial landscape shock in 2022 resulted in more consensus regarding perceptions of Russia in the case countries. This affected regime and niche development in two ways. On the one hand, it created increased support for the expansion of renewable energy. On the other hand, it also formed a stronger consensus about continuing to use fossil fuels in countries where domestic sources were available, especially in Estonia and Norway. The two-pronged impact of the 2022 events means, perhaps, a lack of overall direction for the energy transition. It may also limit the expansion of the energy niches that continue to compete with the fossil fuel-based energy regime.

Viewing a large powerful country as a landscape pressure emphasizes the perspective that, in transition studies, landscape is not merely about physical elements or events but also largely about how landscape factors are perceived by different actors. In this formation of perceptions, that is, socially constructing landscape pressures, regime actors are likely to be more influential than niche actors or other marginal actors. This is well represented in the Finnish case before 2022, where concerns existed but the dominant energy–political logic was that Russia was a partner beneficial to Finland's energy trade and the economy and not a security concern for its energy regime. The view of the landscape being based on perceptions, therefore, connects actors and agency to the conceptualization of the landscape. This has been pointed out previously by Antadze and McGowan (2017), who mentioned how actors interpret the landscape for the use of niches and regimes.

Events and pressures related to large countries also connect to how governments in these countries can put intentional pressure on other actors (cf. Morone et al., 2016). The actions by the Russian state toward Ukraine and the rest of Europe, as well as the sanctions placed by the European Commission in response, represent intentional attempts to create landscape pressure on energy and security regimes. The difference between countries as source of landscape pressure and other landscape developments, such as climate change or pandemics, is that a country can be both an actor via its government and act as a landscape pressure on sociotechnical regimes in other countries.

The country cases also showed how the history and culture of countries, as well as the subculture of particular regimes, shape the ways in which landscape pressures are perceived and interpreted. This is particularly evident in the analysis of Russia as a landscape pressure on the case countries of this book. Estonia had more uniform and risk-oriented perceptions of Russia as a landscape pressure for its energy sector than the other countries, due to its relatively recent regained independence and history as part of the Soviet Union. Likewise, Finland's orientation to not discuss geopolitics around energy was guided by the history of "Finlandization" following World War II (e.g., Arter, 2000). Johnstone and McLeish (2022) describe a similar occurrence where the wider cultural context built from memories and expectations around the potential of another war have an impact on the sociotechnical landscape.

The empirical energy context highlights the complex and multifaceted nature of sustainability transitions, emphasized in the recent literature. Instead of a relatively straightforward transition where niches expand and stabilize to replace an old sociotechnical regime, the real-world empirical context draws attention to reconfiguration and restabilization (Laakso et al., 2020; Sillak and Kanger, 2020). As noted, the aftermath of 2022 saw two somewhat contrary tracks: the restabilization of fossil energy regimes (in Estonia and Norway) and the expansion of niche energy development. Therefore, the country cases do not show a simple regime decline coupled with niche expansion, but rather a reconfiguration of the energy regimes to include both old and new in a new configuration of the sociotechnical system. Scotland may be an exception to this, stating in its energy transition plan from 2023 that "extraction of fossil fuels is not consistent with our climate obligations, [and] is not the right solution to the energy price crisis" (Scottish Government, 2023, p. 97). It, therefore, takes a longer-term approach to positive security involving the phaseout of fossil fuels.

The restabilization of industries benefits from dominant industrial actors' active counteractions to destabilization. For instance, Sillak and Kanger (2020) note that restabilization strategies include reinforcing territorial ties via existing resources and infrastructure, increasing societal embedding by emphasizing established cultural meanings, and reinforcing existing policy–industry alliances. These strategies were observable in relation to the Estonian oil shale and Finnish peat industries. The cases in this book showed also that a large landscape shock may create greater consensus around regime restabilization than previously existed.

This book has highlighted the need to strongly consider security and geopolitics as areas that are increasingly relevant for sustainability transitions research but have typically been ignored. The book's empirical cases showed how security and defence policy influence both niche development and regime decline in the energy sector. A similar finding was made earlier by Kester et al. (2020); they argued that security concerns hinder niche development in the mobility sector. A further argument this book makes is that unfolding sustainability transitions also affect security and defence regimes by changing the technological operation, actor–constellation, and institutional structures of sociotechnical systems. These types of effects should be analyzed or anticipated ex ante where possible and revised during the course of transitions.

As noted, the focus on security also emphasizes the role of the state in sustainability transitions (see Johnstone and Newell, 2018; Silvester and Fisker, 2023). Whereas the role of the state ranges from hindering to advancing transitions, the analyses in this book showed two things relating to security and defence, both sectors that are typically the responsibility of state governments. First, policy incoherence resulting from political incoherence (see Kivimaa, 2022) and the differing values and worldviews of different government ministries mean that the same state can simultaneously advance and hinder a transition. Second, despite security being a responsibility of the state, private sector actors' activities, expertise, and knowledge are vital in complementing states in their duties at the nexus of security and sustainability transitions.

## 9.6 Final Remarks

With this book, I aimed to introduce the fascinating world of security studies and international relations to researchers, energy sector experts, and those interested in sustainability transitions. It can be especially useful to apply certain concepts from security studies, such as securitization, positive and negative security, and referent objects, to transition studies too. In terms of the energy–security nexus, one can conclude that the referent object, that is, that which is to be secured, ranges from the nation state and broader society via the energy system to individual citizens (while the citizen dimension is less commonly explored it is important for positive security and just transitions). The analyses showed that new security concerns can both accelerate niche expansion and slow down regime decline, depending on context. Therefore, uniform conclusions cannot be made about this link.

I also hope with this book to open a new research agenda that brings security studies and geopolitics into the sustainability transitions scholarship. Based on my analysis and previous work touching on this interface, the following research questions arise. First, what are the ways in which security shapes the emergence of new niches and what roles do militaries play in the uptake of new technologies? Some insights have been provided in the energy and mobility contexts (e.g., Kester et al., 2020) but further research would be beneficial both in the context of new sectors and to deepen the analyses of energy and mobility niches. Second, how are established sociotechnical regimes tied to security and the military–industrial complex, and what needs to take place to open such multiregime lock-ins? We have some insights from the UK context (e.g., Johnstone et al., 2017), but new research is needed across the Global North and the Global South. Third, what are the ways in which sustainability transitions are linked to conflicts and peace-building? Again, there is a limited number of studies in selected contexts (Fischhender et al., 2021; Kester and Sovacool, 2017), and more globally encompassing studies are needed. Fourth, how do transitions link to war and the role of the state? Some interesting openings have been made in this regard (Ford and Newell, 2021; Johnstone and McLeish, 2022; Johnstone and Newell, 2018), but further research is needed, especially in contemporary contexts. Finally, the research in this area also needs to connect to positive security and just transitions, to explore the myriad ways in which security and justice are intertwined.

Security studies have presented a question regarding "security from what threats?" Based on the analysis of this book, technical aspects have dominated the thinking around energy system security. Before 2022, an economic understanding of energy security prevailed and the geopolitical dimension was often ignored (Dyer, 2016). It is only recently that increased attention has been paid to, for instance, military and terrorist threats (i.e., human risks) to critical infrastructure. Also, climate change and security-of-materials supply (i.e., nature-based risks) have increased in importance.

Another question posed in security studies has been "security for what values?" The analyses of this book have shown that economic and market-based values have tended to dominate sociotechnical energy regimes, whereas environmental values have mainly been covered via attention to climate change, with biodiversity and the threat of nature loss largely ignored. Hard security values have also often been absent, but have been increasing in magnitude since 2022. Soft security has been similarly absent but has also received increasing attention via just transition efforts and emerging discourse on societal resilience during the last few years.

The analyses of this book showed the policy interface around energy transitions and national security and defence has often been incoherent and pursuits toward coherence have been based on informal institutional coordination and depoliticized settings. Further, the empirical experiences indicate that the political dimension, in achieving or not achieving policy coherence, is vital. Across Europe there was a shift in political frames as a result of the events of 2022, with much more potential for coherence between energy transition and security policies than before, but also the risk of again giving increased support to fossil fuel-based energy regimes. This means that policymakers and other actors need to make conscious and ambitious efforts to improve coherence and integration between energy transition policies and security and defence policies.

The search for resilience, strategic autonomy, and technology sovereignty in the EU, its member states, and elsewhere in Europe is perhaps an example of such efforts. These recent policy developments also connect the practice of sustainability transitions, in particular the EU Green Deal, with debates on security and justice. Policies to advance resilience and strategic autonomy must consider how these impact the advancement of sustainability transitions, not only nationally but globally, and what the implications of such pursuits on global security and justice are. Further research is needed in this area. This also raises the need to coordinate sectoral policies, such as energy or industrial policies, not only with defence policies but also with foreign and development policies – linking to changing energy and climate diplomacy.

The topic of this book is also connected to broader discussions on climate security. The zero-carbon energy transition has an important role to play for future climate security by reducing greenhouse gas emissions. There are, however, also other connections. For instance, new sociotechnical energy systems need to be built so that this critical infrastructure is resilient to the increasing impacts of climate change, such as storms, droughts, flooding, heatwaves, and fires. The energy transition and climate change together create increased pressure on land use and alter global trade and supply chains.

For some time, human-induced climate change has been considered by NATO and large countries' militaries as an existential threat, something that changes the operational capabilities of militaries and that needs to be mainstreamed to the operation of militaries and defence forces. Such attention indicates that the threat is real. There are also examples in the ways in which zero-carbon energy policies and defence policies are becoming more integrated with each other, evidenced, for instance, in a NATO-funded workshop that Chatham House co-organized with the Finnish Environment Institute in September 2023. At the same time, the discussions held in this workshop revealed that much is still to be done: thinking about how more concretely to mainstream climate security in NATO and its member countries; developing alternative technologies and fuels for operations; and considering the justice implications of climate change mitigation and adaptation.

I end by noting that improved policy interplay between energy transition, defence, and security policies requires institutional change. One part of such institutional change should be redefining what energy security means in the context of a new kind of decarbonized energy regime. For instance, energy security in the context of renewable energy and electrification-based transitions can imply securing cross-border electricity interconnections, distributed smart grids, improved electricity storage, and international energy collaboration; preparing for disturbances; tightening public–private cooperation; and establishing new business models around demand response. Energy efficiency was usually not connected with energy security in the case countries, although reduced energy demand would improve security of supply and lessen pressures around sourcing technological components and critical materials. Policymakers and others need to create better links between questions of energy efficiency and of security.

# References

Abernathy, W. J., Clark, K. B., 1985. Innovation: Mapping the winds of creative destruction. *Research Policy* 14, 3–22. https://doi.org/10.1016/0048-7333(85)90021-6

Abraham, J., 2019. Just transitions in a dual labor market: Right wing populism and austerity in the German *energiewende*. *Journal of Labor and Society* 22, 679–693. https://doi.org/10.1111/LANDS.12438

Adams, N. N., Mueller-Hirth, N., 2021. Collaborate and die! Exploring different understandings of organisational cooperation within Scotland's uncertain North Sea oil and gas industry. *Energy Research & Social Science* 73, 101909. https://doi.org/10.1016/j.erss.2021.101909

Afewerki, S., Karlsen, A., 2022. Policy mixes for just sustainable development in regions specialized in carbon-intensive industries: The case of two Norwegian petro-maritime regions. *European Planning Studies* 30, 2273–2292. https://doi.org/10.1080/09654313.2021.1941786

Alberts, E. C., 2023. Norway proposes opening Germany-sized area of its continental shelf to deep-sea mining. *Mongabay*. April 20, 2023. https://news.mongabay.com/2023/04/norway-proposes-opening-germany-sized-area-of-its-continental-shelf-to-deep-sea-mining/

Allenby, B. R., 2016. Environmental security: Concept and implementation. *International Political Science Review* 21, 5–21. https://doi.org/10.1177/0192512100211001

Andersen, A. D., Gulbrandsen, M., 2020. The innovation and industry dynamics of technology phase-out in sustainability transitions: Insights from diversifying petroleum technology suppliers in Norway. *Energy Research & Social Science* 64, 101447. https://doi.org/10.1016/j.erss.2020.101447

Andersson, J., Hellsmark, H., Sandén, B., 2021. The outcomes of directionality: Towards a morphology of sociotechnical systems. *Environmental Innovation and Societal Transitions* 40, 108–131. https://doi.org/10.1016/J.EIST.2021.06.008

Ang, B. W., Choong, W. L., Ng, T. S., 2015. Energy security: Definitions, dimensions and indexes. *Renewable and Sustainable Energy Reviews* 42, 1077–1093. https://doi.org/10.1016/J.RSER.2014.10.064

Antadze, N., McGowan, K. A., 2017. Moral entrepreneurship: Thinking and acting at the landscape level to foster sustainability transitions. *Environmental Innovation and Societal Transitions* 25, 1–13. https://doi.org/10.1016/J.EIST.2016.11.001

Aradau, C., 2004. Security and the democratic scene: Desecuritization and emancipation. *Journal of International Relations and Development* 7, 388–413. https://doi.org/10.1057/palgrave.jird.1800030

Arter, D., 2000. Small state influence within the EU: The case of Finland's "Northern Dimension Initiative." *JCMS: Journal of Common Market Studies* 38, 677–697. https://doi.org/10.1111/1468-5965.00260

Auer, M. R., 1998. Environmentalism and Estonia's independence movement. *Nationalities Papers* 26(4), 659–676. DOI: 10.1080/00905999808408593

Austin, P. L., 2021. This company was hit with a devastating ransomware attack. *TIME*. 07.14.2021. https://time.com/6080293/norsk-hydro-ransomware-attack/

Avelino, F., 2021. Theories of power and social change. Power contestations and their implications for research on social change and innovation. *Journal of Political Power* 1–24. https://doi.org/10.1080/2158379X.2021.1875307

Avelino, F., Wittmayer, J. M., 2015. Shifting power relations in sustainability transitions: A multi-actor perspective. *Journal of Environmental Policy and Planning* 18, 628–649. https://doi.org/10.1080/1523908X.2015.1112259

Azzuni, A., Breyer, C., 2018. Definitions and dimensions of energy security: A literature review. *Wiley Interdisciplinary Reviews: Energy and Environment* 7, e268. https://doi.org/10.1002/WENE.268

Bang, G., Lahn, B., 2020. From oil as welfare to oil as risk? Norwegian petroleum resource governance and climate policy. *Climate Policy* 20, 997–1009. https://doi.org/10.1080/14693062.2019.1692774

Bazilian, M., Sovacool, B., Moss, T., 2017. Rethinking energy statecraft: United States foreign policy and the changing geopolitics of energy. *Global Policy* 8(3), 422–425. https://doi.org/10.1111/1758-5899.12461

BEIS, 2022. Record funding uplift for UK battery research and development. Department for Business, Energy and Industrial Strategy, London, UK.

BEIS, 2023. Resilience for the future: The UK's critical minerals strategy. Department for Business, Energy and Industrial Strategy, London, UK.

Berkhout, F., Angel, D., Wieczorek, A. J., 2009. Asian development pathways and sustainable socio-technical regimes. *Technological Forecasting and Social Change* 76, 218–228. https://doi.org/10.1016/J.TECHFORE.2008.03.017

Berling, T. V., Gad, U. P., Petersen, K. L., Wæver, O., 2021. *Translations of Security: A Framework for the Study of Unwanted Futures*. Routledge, Abingdon and New York. https://doi.org/10.4324/9781003175247

Berzina, I. 2020. From "Total" to "Comprehensive" national defence: The development of the concept in Europe. *Journal on Baltic Security* 6(2), 1–9. Baltic Defence College. DOI: 10.2478/jobs-2020-0006

Blindheim, B., 2013. Implementation of wind power in the Norwegian market; the reason why some of the best wind resources in Europe were not utilised by 2010. *Energy Policy* 58, 337–346. https://doi.org/10.1016/j.enpol.2013.03.033

Blondeel, M., Bradshaw, M. J., Bridge, G., Kuzemko, C., 2021. The geopolitics of energy system transformation: A review. *Geography Compass* 15, e12580. https://doi.org/10.1111/GEC3.12580

Blythe, J., Silver, J., Evans, L., Armitage, D., Bennett, N. J., Moore, M. L., Morrison, T. H., Brown, K., 2018. The dark side of transformation: Latent risks in contemporary sustainability discourse. *Antipode* 50, 1206–1223. https://doi.org/10.1111/ANTI.12405

Boasson, E. L., 2021. Norway: Certificate supporters turning opponents, in: Boasson, E. L., Leiren, M. D., Wettestad, J. (Eds.), *Comparative Renewables Policy: Political, Organizational and European Fields*. Routledge, Abingdon and New York, pp. 193–216.

Bocse, A.-M., 2020. NATO, energy security and institutional change. *European Security* 29, 436–455. https://doi.org/10.1080/09662839.2020.1768072

Bögel, P. M., Upham, P., 2018. Role of psychology in sociotechnical transitions studies: Review in relation to consumption and technology acceptance. *Environmental Innovation and Societal Transitions* 28, 122–136. https://doi.org/10.1016/J.EIST.2018.01.002

Bolton, R., 2021. *Making Energy Markets: The Origins of Electricity Liberalisation in Europe.* Springer, Cham.

Booth, K., 1991. Security and emancipation. *Review of International Studies* 17(4), 313–326. doi:10.1017/S0260210500112033

Booth, K., 2007. *Theory of World Security.* Cambridge University Press, Cambridge.

Busby, J. 2022. *States and Nature: The Effects of Climate Change on Security.* Cambridge University Press, Cambridge, UK.

Buzan, B., Wæver, O., de Wilde, J., 1998. *Security: A New Framework for Analysis.* Lynne Riener, Boulder.

Candel, J. J. L., 2021. The expediency of policy integration. *Policy Studies* 42, 346–361. https://doi.org/10.1080/01442872.2019.1634191

Candel, J. J. L., Biesbroek, R., 2016. Toward a processual understanding of policy integration. *Policy Sciences* 49, 211–231. https://doi.org/10.1007/s11077-016-9248-y

Carbone, M., 2008. Mission impossible: The European Union and policy coherence for development. *Journal of European Integration* 30, 323–342. https://doi.org/10.1080/07036330802144992

Carter, T. R., Benzie, M., Campiglio, E., Carlsen, H., Fronzek, S., Hildén, M., Reyer, C. P. O., West, C., 2021. A conceptual framework for cross-border impacts of climate change. *Global Environmental Change* 69. https://doi.org/10.1016/j.gloenvcha.2021.102307

CCC, 2022. *Progress in Reducing Emissions in Scotland: 2022 Report to Parliament.* Climate Change Committee, London, UK.

Chaar, A. M., Mangalagiu, D., Khoury, A., Nicolas, M., 2020. Transition towards sustainability in a post-conflict country: A neo-institutional perspective on the Lebanese case. *Climatic Change* 160, 691–709. https://doi.org/10.1007/S10584-019-02478-7/TABLES/2

Cherp, A., Jewell, J., 2011. The three perspectives on energy security: Intellectual history, disciplinary roots and the potential for integration. *Current Opinion in Environmental Sustainability* 3(4): 202–212. https://doi.org/10.1016/j.cosust.2011.07.001

Cherp, A., Jewell, J., 2014. The concept of energy security: Beyond the four as. *Energy Policy* 75, 415–421. https://doi.org/10.1016/j.enpol.2014.09.005

Chester, L., 2010. Conceptualising energy security and making explicit its polysemic nature. *Energy Policy* 38, 887–895. https://doi.org/10.1016/J.ENPOL.2009.10.039

Child, M., Kemfert, C., Bogdanov, D., Breyer, C., 2019. Flexible electricity generation, grid exchange and storage for the transition to a 100% renewable energy system in Europe. *Renewable Energy* 139, 80–101. https://doi.org/10.1016/J.RENENE.2019.02.077

Cook, C., Bakker, K., 2012. Water security: Debating an emerging paradigm. *Global Environmental Change* 22, 94–102. https://doi.org/10.1016/J.GLOENVCHA.2011.10.011

Cornell, P., 2019. International grid integration: Efficiencies, vulnerabilities, and strategic implications in Asia. *Atlantic Council.* www.atlanticcouncil.org/in-depth-research-reports/report/international-grid-integration-efficiencies-vulnerabilities-and-strategic-implications-in-asia/

Corry, O., 2011. Securitisation and "Riskification": Second-order security and the politics of climate change. *Millennium: Journal of International Studies* 40, 235–258. https://doi.org/10.1177/0305829811419444

Cortright, D., Seyle, C., Wall, K., 2017. *Governance for Peace.* Cambridge University Press, Cambridge, UK.

Cowell, R., Ellis, G., Sherry-Brennan, F., Strachan, P. A., Toke, D., 2017. Energy transitions, sub-national government and regime flexibility: How has devolution in the United Kingdom affected renewable energy development? *Energy Research & Social Science* 23, 169–181. https://doi.org/10.1016/j.erss.2016.10.006

CPTRA, 2023. Superficies licences for offshore wind farms. Consumer Protection and Technical Regulatory Authority of Estonia. Accessed November 2, 2023: https://ttja.ee/en/business-client/buildings-construction/superficies-licences-offshore-wind-farms

Crandall, M., 2014. Soft security threats and small states: The case of Estonia. *Defence Studies* 14, 30–55. https://doi.org/10.1080/14702436.2014.890334

Criekemans, D., 2018. Geopolitics of the renewable energy game and its potential impact upon global power relations, in: Scholten, D. (Ed.), *Geopolitics of Renewables*. Springer, pp. 37–73. https://doi.org/10.1007/978-3-319-67855-9_2

Dalby, S., 2002. *Environmental Security*. University of Minnesota Press, Minneapolis.

Depledge, D., 2023. Low-carbon warfare: Climate change, net zero and military operations. *International Affairs* iiad001. https://doi.org/10.1093/ia/iiad001

Depledge, D., Kennedy-Pipe, C., Rogers, J., 2019. The UK and the Arctic: Forward defence, in: *Arctic Yearbook 2019*. https://arcticyearbook.com/arctic-yearbook/2019

Desmidt, S., 2021. Climate change and security in North Africa Focus on Algeria, Morocco and Tunisia. Cascades Research Paper. Accessed October 9, 2023: www.cascades.eu/wp-content/uploads/2021/02/CASCADES-Research-paper-Climate-change-and-security-in-North-Africa-1.pdf

DESNZ, 2023a. Energy trends. Accessed October 6, 2023: https://assets.publishing.service.gov.uk/government/uploads/system/uploads/attachment_data/file/1126161/Energy_Trends_December_2022.pdf

DESNZ, 2023b. Digest of UK Energy Statistics (DUKES): Energy. Accessed October 6, 2023: www.gov.uk/government/statistics/energy-chapter-1-digest-of-united-kingdom-energy-statistics-dukes

Dorfman, A., 2017. *The Future of British Defence Policy* (No. 74). IFRI, Paris.

Dorfman, P., 2021. Climate impact: UK nuclear military. Nuclear Consulting Group. Accessed October 9, 2023: www.preventionweb.net/publication/climate-impact-uk-nuclear-military

Dupont, C., 2019. The EU's collective securitisation of climate change. *West European Politics* 42, 369–390. https://doi.org/10.1080/01402382.2018.1510199

Dyer, H., 2016. Energy Security and Energy Policy Incoherence, in: *Delivering Energy Law and Policy in the EU and the US*. Edinburgh University Press, Edinburgh.

Earnst & Young, 2023. Just Transition Review of the Scottish Energy Sector: Summary Report. Accessed March 31, 2023: www.energy-system-and-just-transition-independent-analysis.co.uk/summary-report.pdf

EC, 2014. COMMUNICATION FROM THE COMMISSION TO THE EUROPEAN PARLIAMENT AND THE COUNCIL European Energy Security Strategy, COM(2014) 330 final. European Commission, Brussels, Belgium.

EC, 2020. Critical raw materials resilience: Charting a path towards greater security and sustainability, COMMUNICATION FROM THE COMMISSION TO THE EUROPEAN PARLIAMENT, THE COUNCIL, THE EUROPEAN ECONOMIC AND SOCIAL COMMITTEE AND THE COMMITTEE OF THE REGIONS. European Commission, Brussels, Belgium.

Edmondson, D. L., Kern, F., Rogge, K. S., 2019. The co-evolution of policy mixes and socio-technical systems: Towards a conceptual framework of policy mix feedback in sustainability transitions. *Research Policy* 48. https://doi.org/10.1016/j.respol.2018.03.010

Equinor, 2023. *2022 Integrated Annual Report*. Equinor, Stavanger.
ERR, 2021. Competition authority: Electricity market heavily import-dependent. ERR 3.11.2021. Accessed October 9, 2023: https://news.err.ee/1608390869/competition-authority-electricity-market-heavily-import-dependent
Estonian Defence Forces, 2023. Estonian defence forces. Accessed October 4, 2023: https://mil.ee/en/defence-forces/
Eurostat, 2023. Energy statistics – An overview. Accessed 22.08.2023: https://ec.europa.eu/eurostat/statistics-explained/index.php?title=Energy_statistics_-_an_overview#:~:text=EU%20energy%20import%20dependency%20rate%20stood%20at%2055.5%25%20in%202021.&text=Gross%20available%20energy%20in%20the,by%206.0%25%20compared%20wi
Fankhouser, S., Averchenkova, A., Finnegan, J., 2018. 10 years of the UK Climate Change Act. London School of Economics, London, UK. Accessed March 30, 2023: www.lse.ac.uk/GranthamInstitute/wp-content/uploads/2018/03/10-Years-of-the-UK-Climate-Change-Act_Fankhauser-et-al.pdf
Farham, A., Kossman, S., van Rij, A. 2023. *Preparing NATO for Climate-Related Security Challenges*. Chatham House Research Paper. Chatham House, London.
Feola, G., Vincent, O., Moore, D., 2021. (Un)making in sustainability transformation beyond capitalism. *Global Environmental Change* 69, 102290. https://doi.org/10.1016/j.gloenvcha.2021.102290
Finland's Wind Power Association, 2022. Tuulivoima Suomessa 2022. Accessed October 5, 2023: https://tuulivoimayhdistys.fi/media/tuulivoima_vuositilastot_2022-1.pdf
Finland's Wind Power Association, 2023. Tuulivoimatilastot 6/2023. Accessed October 5, 2023: https://tuulivoimayhdistys.fi/ajankohtaista/tilastot-2/tuulivoiman-rakentamisen-tahti-jatkuu-tasaisena-2
Finnish Defence Forces, 2022. Puolustusvoimien energia- ja ilmasto-ohjelman 2022–2025 tavoitteet ja toimenpiteet. Defence Forces, Helsinki.
Finnish Security Committee, 2017. Security strategy for society, government resolution 2.11.2017. Helsinki.
Fischhendler, I., Herman, L., David, L., 2021. Light at the end of the panel: The Gaza strip and the interplay between geopolitical conflict and renewable energy transition. *New Political Economy* 27(1), 1–18. https://doi.org/10.1080/13563467.2021.1903850
Fischhendler, I., Herman, L., Maoz, N., 2017. The political economy of energy sanctions: Insights from a global outlook 1938–2017. *Energy Research & Social Science* 34, 62–71. https://doi.org/10.1016/j.erss.2017.05.008
Flanagan, K., Uyarra, E., Laranja, M., 2011. Reconceptualising the "policy mix" for innovation. *Research Policy* 40, 702–713. https://doi.org/10.1016/J.RESPOL.2011.02.005
Fleming, C., 2021. What would an independent Scotland's defence and security priorities be? in: Hepburn, E., Keating, M., McEwen, N. (Eds.), *Scotland's New Choice: Independence after Brexit*. Centre on Constitutional Change, Edinburgh, 189–197.
Floyd, R., 2019. *The Morality of Security: A Theory of Just Securitisation*. Cambridge University Press, Cambridge, UK.
Ford, A., Newell, P., 2021. Regime resistance and accommodation: Toward a neo-Gramscian perspective on energy transitions. *Energy Research & Social Science* 79, 102163. https://doi.org/10.1016/J.ERSS.2021.102163
Formin, 2016. Fennovoima Oy:n ydinvoimalahanketta koskeva rakentamislupa-hakemus; ulkoasiainministeriön lausunto 21.6.2023. Accessed October 9, 2023: https://tem.fi/documents/1410877/2616019/Ulkoministeri%C3%B6n+lausunto.pdf

Foxon, T. J., 2011. A coevolutionary framework for analysing a transition to a sustainable low carbon economy. *Ecological Economics* 70, 2258–2267. https://doi.org/10.1016/J.ECOLECON.2011.07.014

Freeman, D., 2018. China and renewables: The priority of economics over geopolitics, in: Scholten, D. (Ed.), *Geopolitics of Renewables*. Springer, pp. 187–201. https://doi.org/10.1007/978-3-319-67855-9_7

Furness, M., Gänzle, S., 2017. The security–development Nexus in European Union foreign relations after Lisbon: Policy coherence at last? *Development Policy Review* 35, 475–492. https://doi.org/10.1111/dpr.12191

Geels, F. W., 2002. Technological transitions as evolutionary reconfiguration processes: A multi-level perspective and a case-study. *Research Policy* 31, 1257–1274. https://doi.org/10.1016/s0048-7333(02)00062-8

Geels, F. W., 2004. From sectoral systems of innovation to socio-technical systems: Insights about dynamics and change from sociology and institutional theory. *Research Policy* 33, 897–920. https://doi.org/10.1016/j.respol.2004.01.015

Geels, F. W., 2005a. Co-evolution of technology and society: The transition in water supply and personal hygiene in the Netherlands (1850–1930) – A case study in multi-level perspective. *Technology in Society* 27, 363–397. https://doi.org/10.1016/J.TECHSOC.2005.04.008

Geels, F. W., 2005b. Processes and patterns in transitions and system innovations: Refining the co-evolutionary multi-level perspective. *Technological Forecasting and Social Change* 72, 681–696. https://doi.org/10.1016/j.techfore.2004.08.014

Geels, F. W., 2006. Multi-level perspective on system innovation: Relevance for industrial transformation? in: Olsthoorn, X., Wieczorek, A. J. (Eds.), *Understanding Industrial Transformation: Views from Different Disciplines*. Springer, Dordrecht, pp. 163–186.

Geels, F. W., 2007. Analysing the breakthrough of rock "n" roll (1930–1970) Multi-regime interaction and reconfiguration in the multi-level perspective. *Technological Forecasting & Social Change* 74, 1411–1431. https://doi.org/10.1016/j.techfore.2006.07.008

Geels, F. W., 2011. The multi-level perspective on sustainability transitions: Responses to seven criticisms. *Environmental Innovation and Societal Transitions*. https://doi.org/10.1016/j.eist.2011.02.002

Geels, F. W., 2014. Regime resistance against low-carbon transitions: Introducing politics and power into the multi-level perspective. *Theory, Culture & Society* 31, 21–40. https://doi.org/10.1177/0263276414531627

Geels, F. W., Kern, F., Fuchs, G., Hinderer, N., Kungl, G., Mylan, J., Neukirch, M., Wassermann, S., 2016. The enactment of socio-technical transition pathways: A reformulated typology and a comparative multi-level analysis of the German and UK low-carbon electricity transitions (1990–2014). *Research Policy* 45, 896–913. https://doi.org/10.1016/J.RESPOL.2016.01.015

Geels, F. W., Schot, J., 2007. Typology of sociotechnical transition pathways. *Research Policy*, 36, 399–417. https://doi.org/10.1016/j.respol.2007.01.003

Geels, F. W., Verhees, B., 2011. Cultural legitimacy and framing struggles in innovation journeys: A cultural-performative perspective and a case study of Dutch nuclear energy (1945–1986). *Technological Forecasting & Social Change* 78, 910–930. https://doi.org/10.1016/j.techfore.2010.12.004

Genus, A., Coles, A.-M., 2008. Rethinking the multi-level perspective of technological transitions. *Research Policy* 37, 1436–1445. https://doi.org/10.1016/j.respol.2008.05.006

Ghosh, B., Kivimaa, P., Ramirez, M., Schot, J., Torrens, J., 2021. Transformative outcomes: Assessing and reorienting experimentation with transformative innovation policy. *Science and Public Policy* 48, 739–756. https://doi.org/10.1093/SCIPOL/SCAB045

Ghosh, B., Schot, J., 2019. Towards a novel regime change framework: Studying mobility transitions in public transport regimes in an Indian megacity. *Energy Research & Social Science* 51, 82–95. https://doi.org/10.1016/J.ERSS.2018.12.001

Gjørv, G. H., 2012. Security by any other name: Negative security, positive security, and a multi-actor security approach. *Review of International Studies* 38, 835–859. https://doi.org/10.1017/S0260210511000751

Godzimirski, J. M., 2022. Protection of critical infrastructure in Norway – Factors, actors and systems. *Security and Defence Quarterly* 39, 45–62. https://doi.org/10.35467/sdq/151964

Gorissen, L., Spira, F., Meynaerts, E., Valkering, P., Frantzeskaki, N., 2018. Moving towards systemic change? Investigating acceleration dynamics of urban sustainability transitions in the Belgian City of Genk. *Journal of Cleaner Production* 173, 171–185. https://doi.org/10.1016/J.JCLEPRO.2016.12.052

Greim, P., Solomon, A. A., Breyer, C., 2020. Assessment of lithium criticality in the global energy transition and addressing policy gaps in transportation. *Nature Communications* 11, 1–11. https://doi.org/10.1038/s41467-020-18402-y

Griffiths, S., 2019. Energy diplomacy in a time of energy transition. *Energy Strategy Reviews* 26, 100386. https://doi.org/10.1016/j.esr.2019.100386

Grin, J., Rotmans, J., Schot, J. W., 2010. *Transitions to Sustainable Development New Directions in the Study of Long Term Transformative Change*. Routledge, New York.

Groves, C., Henwood, K., Pidgeon, N., Cherry, C., Roberts, E., Shirani, F., Thomas, G., 2021. The future is flexible? Exploring expert visions of energy system decarbonisation. *Futures* 130, 102753. https://doi.org/10.1016/j.futures.2021.102753

Hancock, K. J., Sovacool, B. K., 2018. International political economy and renewable energy: Hydroelectric power and the resource curse. *International Studies Review* 20, 615–632. https://doi.org/10.1093/isr/vix058

Hansen, L., 2012. Reconstructing desecuritisation: The normative-political in the Copenhagen School and directions for how to apply it. *Review of International Studies* 38, 525–546. https://doi.org/10.1017/S0260210511000581

Hansen, S. T., Moe, E., 2022. Renewable energy expansion or the preservation of national energy sovereignty? Norwegian renewable energy policy meets resource nationalism. *Political Geography* 99, 102760. https://doi.org/10.1016/j.polgeo.2022.102760

Haukkala, T., 2018. A struggle for change – The formation of a green-transition advocacy coalition in Finland. *Environmental Innovation and Societal Transitions* 27, 146–156. https://doi.org/10.1016/j.eist.2017.12.001

Healy, N., Barry, J., 2017. Politicizing energy justice and energy system transitions: Fossil fuel divestment and a "just transition." *Energy Policy* 108, 451–459. https://doi.org/10.1016/j.enpol.2017.06.014

Hebinck, A., Diercks, G., von Wirth, T., Beers, P. J., Barsties, L., Buchel, S., Greer, R., van Steenbergen, F., Loorbach, D., 2022. An actionable understanding of societal transitions: The X-curve framework. *Sustainability Science* 1, 1–13. https://doi.org/10.1007/S11625-021-01084-W/FIGURES/3

Heffron, R. J., Nuttall, W. J., 2017. Scotland, nuclear energy policy and independence, in: Wood, G., Baker, K. (Eds.), *A Critical Review of Scottish Renewable and Low Carbon Energy Policy*. Palgrave Macmillan, Cham, Switzerland, pp. 103–126.

Heinrich, A., 2018. Securitisation in the gas sector: Energy security debates concerning the example of the Nord Stream Pipeline, in: Szulecki, K. (Ed.), *Energy Security in Europe*. Palgrave Macmillan, Cham, Switzerland, pp. 61–91. https://doi.org/10.1007/978-3-319-64964-1_3

Heinrich, A., Szulecki, K., 2018. Energy securitisation: Applying the Copenhagen school's framework to energy, in: Szulecki, K. (Ed.), *Energy Security in Europe: Divergent Perceptions and Policy Challenges*. Palgrave Macmillan, Cham, Switzerland, pp. 33–60.

Hillion, C., 2019. Norway and the changing Common Foreign and Security Policy of the European Union, NUPI Report 1/2019. Norwegian Institute of International Affairs, Oslo. Accessed October 10, 2023: https://nupi.brage.unit.no/nupi-xmlui/bitstream/handle/11250/2582455/NUPI_Report_1_2019_Hillion.pdf.

HM Government, 2020. Energy white paper: Powering our net zero future. Her Majesty's Government, London, UK. Accessed May 14, 2024. www.gov.uk/government/publications/energy-white-paper-powering-our-net-zero-future

HM Government, 2021. Global Britain in a competitive age: The integrated review of security, defence, development and foreign policy (Policy Paper No. CP403). HM Government, UK. Accessed October 10, 2023: www.gov.uk/government/publications/global-britain-in-a-competitive-age-the-integrated-review-of-security-defence-development-and-foreign-policy

HM Government, 2023. Integrated review refresh 2023: Responding to a more contested and volatile world. His Majesty's Government, London, UK. Accessed October 10, 2023: www.gov.uk/government/publications/integrated-review-refresh-2023-responding-to-a-more-contested-and-volatile-world

Holmberg, R., 2008. Survival of the unfit: Path dependence and the Estonian oil shale industry. Linköping Studies in Arts and Science No. 427. Doctoral thesis, Linköping University, Sweden.

Holmgren, S., Pever, M., Fischer, K., 2019. Constructing low-carbon futures? Competing storylines in the Estonian energy sector's translation of EU energy goals. *Energy Policy* 135, 111063. https://doi.org/10.1016/j.enpol.2019.111063

Hoogensen Gjørv, G., 2011. Gender, identity, and human security: Can we learn anything from the case of women terrorists? *Canadian Foreign Policy Journal* 12, 119–140. https://doi.org/10.1080/11926422.2005.9673392

Hoogensen Gjørv, G., 2012. Security by any other name: Negative security, positive security, and a multi-actor security approach. *Review of International Studies* 38, 835–859. https://doi.org/10.1017/S0260210511000751

Hoogensen Gjørv, G., Bilgic, A., 2022. *Positive Security: Collective Life in an Uncertain World*. Routledge, Abingdon, Oxon.

Hoogma, R., Kemp, R., Schot, J., Truffer, B., 2002. *Experimenting for Sustainable Transport: The Approach of Strategic Niche Management*. Spon Press, London.

Howlett, M., How, Y., Del Rio, P., 2015. The parameters of policy portfolios: Verticality and horizontality in design spaces and their consequences for policy mix formulation. *Environment and Planning C: Government and Policy* 33. https://doi.org/10.1177/0263774X15610059

Höysniemi, S., 2022. Energy futures reimagined: The global energy transition and dependence on Russian energy as issues in the sociotechnical imaginaries of energy security in Finland. *Energy Research & Social Science* 93, 102840. https://doi.org/10.1016/j.erss.2022.102840

Huda, M. S., 2020. *Energy Cooperation in South Asia: Utilising Natural Resources for Peace and Sustainable Development*. Routledge, New York & Abingdon.

Huttunen, S., Kaljonen, M., Lonkila, A., Rantala, S., Rekola, A., Paloniemi, R., 2021. Pluralising agency to understand behaviour change in sustainability transitions. *Energy Research and Social Science* 76, 102067. https://doi.org/10.1016/j.erss.2021.102067

Huttunen, S., Kivimaa, P., Virkamäki, V., 2014. The need for policy coherence to trigger a transition to biogas production. *Environmental Innovation and Societal Transitions* 12: 14–30. https://doi.org/10.1016/j.eist.2014.04.002

Huysmans, J., 1998. Security! What do you mean?: From concept to thick signifier. *European Journal of International Relations* 4(2), 226–255.

IEA, 2018. Finland 2018 Review. International Energy Agency, Paris. Accessed October 10, 2023: www.iea.org/reports/energy-policies-of-iea-countries-finland-2018-review

IEA, 2019. Estonia 2019 Review. International Energy Agency, Paris. Accessed: May 14, 2024 www.iea.org/reports/energy-policies-of-iea-countries-estonia-2019-review

IEA, 2021. The role of critical minerals in clean energy transitions. International Energy Agency, Paris. Accessed October 10, 2023: www.iea.org/reports/the-role-of-critical-minerals-in-clean-energy-transitions

IEA, 2022. Norway 2022: Energy policy review. International Energy Agency, Paris. Accessed October 10, 2023: www.iea.org/reports/norway-2022

IEA, 2023. Renewable energy. International Energy Agency, Paris. Accessed December 11, 2023: www.iea.org/reports/renewables-2022/renewable-electricity

IEA, 2024. About: Mission statement. Accessed 30 April 2024: www.iea.org/about

Isoaho, K., Markard, J., 2020. The politics of technology decline: Discursive struggles over coal phase-out in the UK. *Review of Policy Research*. https://doi.org/10.1111/ropr.12370

Jääskeläinen, J. J., Höysniemi, S., Syri, S., Tynkkynen, V. P., 2018. Finland's dependence on Russian energy-mutually beneficial trade relations or an energy security threat? *Sustainability (Switzerland)* 10(10): 3445. https://doi.org/10.3390/su10103445

Jenkins, K., Sovacool, B. K., McCauley, D., 2018. Humanizing sociotechnical transitions through energy justice: An ethical framework for global transformative change. *Energy Policy* 117, 66–74. https://doi.org/10.1016/j.enpol.2018.02.036

Jewell, J., Brutschin, E., 2021. The politics of energy security, in: Hancock, K. J., Allison, J. E. (Eds.), *The Oxford Handbook of Energy Politics*. Oxford University Press, Oxford, pp. 247–274. https://doi.org/10.1093/oxfordhb/9780190861360.013.10

Joensuu, K., Väyrynen, L., Tolppanen, J., Karhu, L., Salmi, T., Hartikka, S., Leino, L., Viljanen, J., Smids, S., Hujanen, A., Sipilä, M., Huuskonen, A., 2021. Tuulivoimarakentamisen edistäminen: Keinoja sujuvaan hankekehitykseen ja eri tavoitteiden yhteensovitukseen. Publications of the Prime Minister's Office 2021/51, Helsinki, Finland. Accessed October 10, 2023: https://julkaisut.valtioneuvosto.fi/handle/10024/163302

Johansson, B., 2013. Security aspects of future renewable energy systems – A short overview. *Energy* 61, 598–605. https://doi.org/10.1016/j.energy.2013.09.023

Johnstone, P., Kivimaa, P., 2018. Multiple dimensions of disruption, energy transitions and industrial policy. *Energy Research and Social Science* 37, 260–265. https://doi.org/10.1016/j.erss.2017.10.027

Johnstone, P., McLeish, C., 2022. World wars and sociotechnical change in energy, food, and transport: A deep transitions perspective. *Technological Forecasting and Social Change* 174, 121206. https://doi.org/10.1016/J.TECHFORE.2021.121206

Johnstone, P., Newell, P., 2018. Sustainability transitions and the state. *Environmental Innovation and Societal Transitions* 27, 72–82. https://doi.org/10.1016/J.EIST.2017.10.006

Johnstone, P., Rogge, K. S., Kivimaa, P., Farné Fratini, C., Primmer, E., 2021. Exploring the re-emergence of industrial policy: Perceptions regarding low-carbon energy transitions in Germany, the United Kingdom and Denmark. *Energy Research & Social Science* 74, 101889. https://doi.org/10.1016/j.erss.2020.101889

Johnstone, P., Rogge, K. S., Kivimaa, P., Fratini, C. F., Primmer, E., Stirling, A., 2020. Waves of disruption in clean energy transitions: Sociotechnical dimensions of system disruption in Germany and the United Kingdom. *Energy Research and Social Science* 59, 101287. https://doi.org/10.1016/j.erss.2019.101287

Johnstone, P., Stirling, A., 2020. Comparing nuclear trajectories in Germany and the United Kingdom: From regimes to democracies in sociotechnical transitions and discontinuities. *Energy Research & Social Science* 59, 101245. https://doi.org/10.1016/J.ERSS.2019.101245

Johnstone, P., Stirling, A., Sovacool, B., 2017. Policy mixes for incumbency: Exploring the destructive recreation of renewable energy, shale gas "fracking," and nuclear power in the United Kingdom. *Energy Research and Social Science* 33, 147–162. https://doi.org/10.1016/j.erss.2017.09.005

Jordan, A., Lenschow, A., 2010. Policy paper environmental policy integration: A state of the art review. *Environmental Policy and Governance* 20, 147–158. https://doi.org/10.1002/eet.539

Jordan, M., Hewitt, A., 2022. Depoliticization, participation and social art practice: On the function of social art practice for politicization. *Art & the Public Sphere* 11, 19–36. https://doi.org/10.1386/aps_00066_1

Juozaitis, J., 2020. The synchronization of the Baltic States': Geopolitical implications on the Baltic Sea region and beyond. Energy Highlights. Nato Energy Security Centre of Excellence 1–17. Accessed October 10, 2023: https://enseccoe.org/data/public/uploads/2021/10/d1_baltic-states-synchronization-with-continental-european-network-navigating-the-hybrid-threat-landscape.pdf

Kainiemi, L., Karhunmaa, K., Eloneva, S., 2020. Renovation realities: Actors, institutional work and the struggle to transform Finnish energy policy. *Energy Research & Social Science* 70, 101778. https://doi.org/10.1016/j.erss.2020.101778

Kalantzakos, S., Overland, I., Vakulchuk, R., 2023. Decarbonisation and critical materials in the context of fraught geopolitics: Europe's distinctive approach to a net zero future. *The International Spectator* 58, 3–22. https://doi.org/10.1080/03932729.2022.2157090

Kaljonen, M., Kortetmäki, T., Tribaldos, T., Huttunen, S., Karttunen, K., Maluf, R. S., Niemi, J., Saarinen, M., Salminen, J., Vaalavuo, M., Valsta, L., 2021. Justice in transitions: Widening considerations of justice in dietary transition. *Environmental Innovation and Societal Transitions* 40, 474–485. https://doi.org/10.1016/J.EIST.2021.10.007

Kama, K., 2016. Contending geo-logics: Energy security, resource ontologies, and the politics of expert knowledge in Estonia. *Geopolitics* 21, 831–856. https://doi.org/10.1080/14650045.2016.1210129

Kanger, L., Sovacool, B. K., Noorkõiv, M., 2020. Six policy intervention points for sustainability transitions: A conceptual framework and a systematic literature review. *Research Policy* 49, 104072. https://doi.org/10.1016/j.respol.2020.104072

Karlstrøm, H., Ryghaug, M., 2014. Public attitudes towards renewable energy technologies in Norway. The role of party preferences. *Energy Policy* 67, 656–663. https://doi.org/10.1016/j.enpol.2013.11.049

Kelly, P., 2006. A critique of critical geopolitics. Critique of critical geopolitics. *Geopolitics* 11, 24–53. https://doi.org/10.1080/14650040500524053

Kemp, R., Schot, J., Hoogma, R., 1998. Regime shifts to sustainability through processes of niche formation: The approach of strategic niche management. *Technology Analysis and Strategic Management* 10, 175–198. https://doi.org/10.1080/09537329808524310

Kern, F., Howlett, M., 2009. Implementing transition management as policy reforms: A case study of the Dutch energy sector. *Policy Sciences* 42, 391. https://doi.org/10.1007/s11077-009-9099-x

Kern, F., Kivimaa, P., Martiskainen, M., 2017. Policy packaging or policy patching? The development of complex energy efficiency policy mixes. *Energy Research and Social Science* 23, 11–25. https://doi.org/10.1016/j.erss.2016.11.002

Kester, J., Sovacool, B., 2017. Torn between war and peace: Critiquing the use of war to mobilize peaceful climate action. *Energy Policy* 104, 50–55. https://doi.org/10.1016/j.enpol.2017.01.026

Kester, J., Sovacool, B. K., Noel, L., Zarazua de Rubens, G., 2020. Between hope, hype, and hell: Electric mobility and the interplay of fear and desire in sustainability transitions. *Environmental Innovation and Societal Transitions* 35, 88–102. https://doi.org/10.1016/j.eist.2020.02.004

Kivimaa, P., 2008. Finland: Big is beautiful – Promoting bioenergy in regional-industrial contexts, in: Lafferty, W. M., Ruud, A. (Eds.), *Promoting Sustainable Electricity in Europe: Challenging the Path Dependence of Dominant Energy Systems*. Edward Elgar Publishing, Cheltenham, UK, pp. 159–188.

Kivimaa, P., 2022a. Policy and political (in)coherence, security and Nordic–Baltic energy transitions. *Oxford Open Energy* 1, oiac009. https://doi.org/10.1093/ooenergy/oiac009

Kivimaa, P., 2022b. Transforming innovation policy in the context of global security. *Environmental Innovation and Societal Transitions* 43, 55–61. https://doi.org/10.1016/J.EIST.2022.03.005

Kivimaa, P., Boon, W., Hyysalo, S., Klerkx, L., 2019. Towards a typology of intermediaries in sustainability transitions: A systematic review and a research agenda. *Research Policy* 48(4), 1062–1075. https://doi.org/10.1016/j.respol.2018.10.006

Kivimaa, P., Brisbois, M. C., Jayaram, D., Hakala, E., Siddi, M., 2022. A socio-technical lens on security in sustainability transitions: Future expectations for positive and negative security. *Futures* 141, 102971. https://doi.org/10.1016/J.FUTURES.2022.102971

Kivimaa, P., Kangas, H.-L., Lazarevic, D., 2017. Client-oriented evaluation of "creative destruction" in policy mixes: Finnish policies on building energy efficiency transition. *Energy Research and Social Science* 33, 115–127. https://doi.org/10.1016/j.erss.2017.09.002

Kivimaa, P., Kern, F., 2016. Creative destruction or mere niche support? Innovation policy mixes for sustainability transitions. *Research Policy* 45, 205–217. https://doi.org/10.1016/j.respol.2015.09.008

Kivimaa, P., Laakso, S., Lonkila, A., Kaljonen, M., 2021. Moving beyond disruptive innovation: A review of disruption in sustainability transitions. *Environmental Innovation and Societal Transitions* 38, 110–126. https://doi.org/10.1016/j.eist.2020.12.001

Kivimaa, P., Martiskainen, M., 2018. Dynamics of policy change and intermediation: The arduous transition towards low-energy homes in the United Kingdom. *Energy Research & Social Science* 44, 83–99. https://doi.org/10.1016/J.ERSS.2018.04.032

Kivimaa, P., Rogge, K. S., 2022. Interplay of policy experimentation and institutional change in sustainability transitions: The case of mobility as a service in Finland. *Research Policy* 51, 104412. https://doi.org/10.1016/J.RESPOL.2021.104412

Kivimaa, P., Sivonen, M. H., 2021. Interplay between low-carbon energy transitions and national security: An analysis of policy integration and coherence in Estonia, Finland and Scotland. *Energy Research and Social Science* 75, 102024. https://doi.org/10.1016/j.erss.2021.102024

Kivimaa, P., Sivonen, M. H., 2023. How will renewables expansion and hydrocarbon decline impact security? Analysis from a socio-technical transitions perspective. *Environmental Innovation and Societal Transitions* 48, 100744. https://doi.org/10.1016/j.eist.2023.100744

Knox-Hayes, J., Brown, M. A., Sovacool, B. K., Wang, Y., 2013. Understanding attitudes toward energy security: Results of a cross-national survey. *Global Environmental Change* 23, 609–622. https://doi.org/10.1016/j.gloenvcha.2013.02.003

Koch, N., Tynkkynen, V.-P., 2021. The geopolitics of renewables in Kazakhstan and Russia. *Geopolitics* 26, 521–540. https://doi.org/10.1080/14650045.2019.1583214

Koese, M., Blanco, C. F., Breeman, G., Vijver, M. G., 2022. Towards a more resource-efficient solar future in the EU: An actor-centered approach. *Environmental Innovation and Societal Transitions* 45, 36–51. https://doi.org/10.1016/j.eist.2022.09.001

Köhler, J., Geels, F. W., Kern, F., Markard, J., Onsongo, E., Wieczorek, A., Alkemade, F., Avelino, F., Bergek, A., Boons, F., Fünfschilling, L., Hess, D., Holtz, G., Hyysalo, S., Jenkins, K., Kivimaa, P., Martiskainen, M., McMeekin, A., Mühlemeier, M. S., Nykvist, B., Pel, B., Raven, R., Rohracher, H., Sandén, B., Schot, J., Sovacool, B., Turnheim, B., Welch, D., Wells, P., 2019. An agenda for sustainability transitions research: State of the art and future directions. *Environmental Innovation and Societal Transitions* 31, 1–32. https://doi.org/10.1016/j.eist.2019.01.004

Koppel, K., 2022. Sõnajalgs hoping to bring two Aidu wind turbines online soon. *ERR News* 2.12.2022. Accessed August 25, 2023: https://news.err.ee/1608812890/sonajalgs-hoping-to-bring-two-aidu-wind-turbines-online-soon

Koretsky, Z., Stegmaier, P., Turnheim, B., Lente, H. van (Eds.), 2022. *Technologies in Decline: Socio-Technical Approaches to Discontinuation and Destabilisation*. Routledge, London. https://doi.org/10.4324/9781003213642

Korsnes, M., Loewen, B., Dale, R. F., Steen, M., Skjølsvold, T. M., 2023. Paradoxes of Norway's energy transition: Controversies and justice. *Climate Policy*, available online. https://doi.org/10.1080/14693062.2023.2169238

Kruyt, B., van Vuuren, D. P., de Vries, H. J. M., Groenenberg, H., 2009. Indicators for energy security. *Energy Policy* 37, 2166–2181. https://doi.org/10.1016/J.ENPOL.2009.02.006

Kuus, M., 2017. *Critical Geopolitics*, International Studies. Oxford Research Encyclopedias. https://doi.org/10.1093/acrefore/9780190846626.013.137

Kuzemko, C., 2014. Politicising UK energy: What "speaking energy security" can do. *Policy and Politics* 42, 259–274. https://doi.org/10.1332/030557312X655990

Kuzemko, C., 2016. Energy depoliticisation in the UK: Destroying political capacity. *The British Journal of Politics and International Relations* 18, 107–124. https://doi.org/10.1111/1467-856X.12068

Kuzemko, C., Blondeel, M., Dupont, C., Brisbois, M. C., 2022. Russia's war on Ukraine, European energy policy responses & implications for sustainable transformations. *Energy Research & Social Science* 93, 102842. https://doi.org/10.1016/j.erss.2022.102842

Kuzemko, C., Keating, M., Goldthau, A., 2016. *The Global Energy Challenge: Environment, Development and Security*. Palgrave Macmillan UK, London.

Laakso, S., Aro, R., Heiskanen, E., Kaljonen, M., 2020. Reconfigurations in sustainability transitions: A systematic and critical review. *Sustainability: Science, Practice and Policy* 17, 15–31. https://doi.org/10.1080/15487733.2020.1836921

Lafferty, W., Hovden, E., 2003. Environmental policy integration: Towards an analytical framework. *Environmental Politics* 12, 1–22. https://doi.org/10.1080/09644010412 331308254

Lawhon, M., Silver, J., Ernstson, H., Pierce, J., 2016. Unlearning (Un)located ideas in the provincialization of urban theory. *Regional Studies* 50(9), 1611–1622. https://doi.org/10.1080/00343404.2016.11622

Lawson, A., 2022. UK's first lithium refinery to be built in boost for electric car industry. *The Guardian* 7.11.2022. Accessed October 10, 2023: www.theguardian.com/business/2022/nov/07/uk-lithium-refinery-electric-car-industry-green-lithium-pd-ports-teesport

Lederer, M., 2022. The promise of Prometheus and the opening up of Pandora's Box: Anthropological geopolitics of renewable energy. *Geopolitics* 27, 655–679. https://doi.org/10.1080/14650045.2020.1820486

Lee, J., Bazilian, M., Sovacool, B., Hund, K., Jowitt, S. M., Nguyen, T. P., Månberger, A., Kah, M., Greene, S., Galeazzi, C., Awuah-Offei, K., Moats, M., Tilton, J., Kukoda, S., 2020. Reviewing the material and metal security of low-carbon energy transitions. *Renewable and Sustainable Energy Reviews* 124, 109789. https://doi.org/10.1016/j.rser.2020.109789

Lempinen, H., 2019. "Barely surviving on a pile of gold": Arguing for the case of peat energy in 2010s Finland. *Energy Policy* 128, 1–7. https://doi.org/10.1016/J.ENPOL.2018.12.041

Leonard, A., Ahsan, A., Charbonnier, F., Hirmer, S., 2022. The resource curse in renewable energy: A framework for risk assessment. *Energy Strategy Reviews* 41, 100841. https://doi.org/10.1016/j.esr.2022.100841

Lockwood, M., 2021. A hard act to follow? The evolution and performance of UK climate governance. *Environmental Politics* 30, 26–48. https://doi.org/10.1080/09644016.2021.1910434

Lockwood, M., Kuzemko, C., Mitchell, C., Hoggett, R., 2017. Historical institutionalism and the politics of sustainable energy transitions: A research agenda. *Environment and Planning C: Politics and Space* 35, 312–333. https://doi.org/10.1177/0263774X16660561

Lockwood, M., Mitchell, C., Hoggett, R., 2022. Energy governance in the United Kingdom, in: Knodt, M., Kemmerzell, J. (Eds.), *Handbook of Energy Governance in Europe*. Springer, Cham, pp. 1255–1285.

Lubell, H., 1961. Security of supply and energy policy in Western Europe. *World Politics* 13, 400–422. https://doi.org/10.2307/2009482

Lundvall, B. Å., Borrás, S., 2005. Science, technology and innovation policy, in: Fagerberg, J., Movery, D. C. (Eds.), *The Oxford Handbook of Innovation*, Oxford University Press, Oxford, pp. 599–631.

MacKinnon, D., Karlsen, A., Dawley, S., Steen, M., Afewerki, S., Kenzhegaliyeva, A., 2022. Legitimation, institutions and regional path creation: A cross-national study of offshore wind. *Regional Studies* 56, 644–655. https://doi.org/10.1080/00343404.2020.1861239

MacNeil, R., Beauman, M., 2022. Understanding resistance to just transition ideas in Australian coal communities. *Environmental Innovation and Societal Transitions* 43, 118–126. https://doi.org/10.1016/J.EIST.2022.03.007

Mäkitie, T., 2020. Corporate entrepreneurship and sustainability transitions Resource redeployment of oil and gas industry firms in floating wind power. *Technology Analysis*

& Strategic Management 32, 474–488. https://doi.org/10.1080/09537325.2019.1668553

Mäkitie, T., Normann, H. E., Thune, T. M., Sraml Gonzalez, J., 2019. The green flings: Norwegian oil and gas industry's engagement in offshore wind power. *Energy Policy* 127, 269–279. https://doi.org/10.1016/j.enpol.2018.12.015

Månberger, A., Johansson, B., 2019. The geopolitics of metals and metalloids used for the renewable energy transition. *Energy Strategy Reviews* 26, 100394. https://doi.org/10.1016/j.esr.2019.100394

Månsson, A., Johansson, B., Nilsson, L. J., 2014. Assessing energy security: An overview of commonly used methodologies. *Energy* 73, 1–14. https://doi.org/10.1016/J.ENERGY.2014.06.073

Marín, A., Goya, D., 2021. Mining – The dark side of the energy transition. *Environmental Innovation and Societal Transitions* 41, 86–88. https://doi.org/10.1016/J.EIST.2021.09.011

Markard, J., 2020. The life cycle of technological innovation systems. *Technological Forecasting and Social Change* 153. https://doi.org/10.1016/j.techfore.2018.07.045

Markard, J., Geels, F. W., Raven, R., 2020. Challenges in the acceleration of sustainability transitions. *Environmental Research Letters* 15, 081001. https://doi.org/10.1088/1748-9326/AB9468

Markard, J., Raven, R., Truffer, B., 2012. Sustainability transitions: An emerging field of research and its prospects. *Research Policy* 41, 955–967. https://doi.org/10.1016/j.respol.2012.02.013

Mata Pérez, M. de la E., Scholten, D., Smith Stegen, K., 2019. The multi-speed energy transition in Europe: Opportunities and challenges for EU energy security. *Energy Strategy Reviews* 26, 100415. https://doi.org/10.1016/j.esr.2019.100415

May, P. J., Sapotichne, J., Workman, S., 2006. Policy coherence and policy domains. *Policy Studies Journal* 34, 381–403. https://doi.org/10.1111/j.1541-0072.2006.00178.x

McCulloch, N., Natalini, D., Hossain, N., Justino, P., 2022. An exploration of the association between fuel subsidies and fuel riots. *World Development* 157, 105935. https://doi.org/10.1016/j.worlddev.2022.105935

McSweeney, B., 1999. *Security, Identity and Interests: A Sociology of International Relations*. Cambridge University Press, New York.

MEAC, 2022. Gas market. Ministry of Economic Affairs and Communications, Tallinn, Estonia. Accessed March 31, 2022: www.mkm.ee/en/objectives-activities/energy-sector/gas-market.

MEE, 2014. Tuulivoiman edistämistyöryhmän loppuraportti. Publications of the Ministry of Economic Affairs and Employment 3/2014. Accessed October 10, 2023: https://tem.fi/documents/1410877/2859687/Tuulivoiman+edist%C3%A4misty%C3%B6ryhm%C3%A4n+loppuraportti+31032014.pdf

MEE, 2017. Government report on the National Energy and Climate Strategy for 2030. Publications of the Ministry of Economic Affairs and Employment 12/2027, Helsinki, Finland. Accessed October 10, 2023: https://tem.fi/documents/1410877/2769658/Government+report+on+the+National+Energy+and+Climate+Strategy+for+2030/0bb2a7be-d3c2-4149-a4c2-78449ceb1976

MEE, 2022. Carbon neutral Finland 2035 – National climate and energy strategy. Publications of the Ministry of Economy Affairs and Employment 2022:55, Helsinki, Finland. Accessed October 10, 2023: https://julkaisut.valtioneuvosto.fi/bitstream/handle/10024/164323/TEM_2022_55.pdf

Michaux, S. P., 2021. Assessment of the extra capacity required of alternative energy electrical power systems to completely replace fossil fuels. GTK Open File Work Report 42/2021. Espoo.

Mickwitz, P., Aix, F., Beck, S., Carrs, D., Ferrand, N., Görg, C., Jenssen, A., Kivimaa, P., Kuhlicke, C., Kuindersma, W., Manez, M., Melanen, M., Monni, S., Pedersen, A. B., Reinert, H., van Bommel, S., 2009. Climate policy integration, coherence and governance. PEER Report No. 2. Partnership for European Environmental Research, Helsinki. Accessed October 10, 2023: https://pure.au.dk/ws/files/56076592/PEER_Report2.pdf

Ministry of Defence, 2019. Ministry of defence single departmental plan. Ministry of Defence, London UK. Accessed June 7, 2023: www.gov.uk/government/publications/ministry-of-defence-single-departmental-plan/ministry-of-defence-single-departmental-plan-2019

Ministry of Defence, 2021. Public opinion on national defence 2021. Estonian Ministry of Defence, Tallinn, Estonia. Accessed October 27, 2023: https://kaitseministeerium.ee/sites/default/files/elfinder/article_files/public_opinion_and_national_defence_2021_spring.pdf

MoD, 2015. Sustainable MOD strategy: Act & evolve. 2015–2025. Accessed October 10, 2023: https://assets.publishing.service.gov.uk/media/5a803debed915d74e622d4d5/Sustainable_MOD_Strategy_2015-2025.pdf

MoD, 2021. Ministry of defence climate change and sustainability strategic approach. Ministry of Defence, London, UK. Accessed: May 14, 2024. www.gov.uk/government/publications/ministry-of-defence-climate-change-and-sustainability-strategic-approach

MoD, 2023. Puolustusmenot. Ministry of Defence, Helsinki, Finland. Accessed October 5, 2023: www.defmin.fi/ministerio/toiminta_ja_talous/puolustusmenot#5316504d

MoFA, 2020. Government report on Finnish foreign and security policy, 2020:32. Ministry for Foreign Affairs of Finland, Helsinki, Finland. Accessed October 10, 2023: https://julkaisut.valtioneuvosto.fi/handle/10024/162515

Morone, P., Lopolito, A., Anguilano, D., Sica, E., Tartiu, V. E., 2016. Unpacking landscape pressures on socio-technical regimes: Insights on the urban waste management system. *Environmental Innovation and Societal Transitions* 20, 62–74. https://doi.org/10.1016/J.EIST.2015.10.005

Mowery, D. C., 2012. Defense-related R&D as a model for "grand Challenges" technology policies. *Research Policy* 41, 1703–1715. https://doi.org/10.1016/j.respol.2012.03.027

Munro, F., 2018. Renewable energy in Scotland: Extending the transition-periphery dynamics approach. PhD thesis, University of Glasgow, Glasgow. https://theses.gla.ac.uk/8714/

Musiolik, J., Markard, J., Hekkert, M., 2012. Networks and network resources in technological innovation systems: Towards a conceptual framework for system building. *Technological Forecasting and Social Change* 79, 1032–1048. https://doi.org/10.1016/J.TECHFORE.2012.01.003

Myllyntaus, T., 1991. *Electrifying Finland: The Transfer of a New Technology into a Late Industrialising Economy*. Macmillan, London, UK.

Naber, R., Raven, R., Kouw, M., Dassen, T., 2017. Scaling up sustainable energy innovations. *Energy Policy* 110, 342–354. https://doi.org/10.1016/j.enpol.2017.07.056

Neal, A. W., 2017. *Security in a Small Nation: Scotland, Democracy, Politics*. Open Book Publishers, Cambridge. Accessed October 10, 2023: www.openbookpublishers.com/books/10.11647/obp.0078

Nemes, G., Chiffoleau, Y., Zollet, S., Collison, M., Benedek, Z., Colantuono, F., Dulsrud, A., Fiore, M., Holtkamp, C., Kim, T.-Y., Korzun, M., Mesa-Manzano, R., Reckinger, R., Ruiz-Martínez, I., Smith, K., Tamura, N., Viteri, M. L., Orbán, É. 2021. The impact of COVID-19 on alternative and local food systems and the potential for the sustainability transition: Insights from 13 countries. *Sustainable Production and Consumption* 28, 591–599. https://doi.org/10.1016/j.spc.2021.06.022

NESA, 2022. Vuosikatsaus/Annual report 2021. National Emergency Supply Agency, Helsinki, Finland. Accessed October 10, 2023: www.huoltovarmuuskeskus.fi/files/72ebfbbd7bb98ee70a14437fd291ab29638e854b/hvk-vuosikatsaus-2021.pdf

NESA, 2023. The national emergency supply agency. National Emergency Supply Agency, Helsinki, Finland. Accessed October 10, 2023: www.huoltovarmuuskeskus.fi/en/organisation/the-national-emergency-supply-agency

Nilsson, M., Persson, Å., 2003. Framework for analysing environmental policy integration. *Journal of Environmental Policy and Planning* 5, 333–359. https://doi.org/10.1080/1523908032000171648

Nilsson, M., Zamparutti, T., Petersen, J. E., Nykvist, B., Rudberg, P., McGuinn, J., 2012. Understanding policy coherence: Analytical framework and examples of sector-environment policy interactions in the EU. *Environmental Policy and Governance* 22, 395–423. https://doi.org/10.1002/eet.1589

Nokkala, Arto, 2014. *Kyky ja tahto: Suomen puolustus murroksessa, Suomen puolustus murroksessa*. Docendo, Jyväskylä.

Normann, S., 2021. Green colonialism in the Nordic context: Exploring Southern Saami representations of wind energy development. *Journal of Community Psychology* 49, 77–94. https://doi.org/10.1002/jcop.22422

Norwegian Ministry of Foreign Affairs, 2019. Meld. St. 27 (2018–2019) Report to the Storting (white paper) Norway's role and interests in multilateral cooperation. Norwegian Ministry of Foreign Affairs, Oslo.

Novalia, W., McGrail, S., Rogers, B. C., Raven, R., Brown, R. R., Loorbach, D., 2022. Exploring the interplay between technological decline and deinstitutionalisation in sustainability transitions. *Technological Forecasting and Social Change* 180, 121703. https://doi.org/10.1016/j.techfore.2022.121703

Nyberg, D., Wright, C., Kirk, J., 2018. Dash for gas: Climate change, hegemony and the scalar politics of fracking in the UK. *British Journal of Management* 29, 235–251. https://doi.org/10.1111/1467-8551.12291

Oesch, P., 2022. Energiahuoltovarmuus testissä. *Maanpuolustus* 142, 20–23. Accessed October 10, 2023: www.maanpuolustus-lehti.fi/energiahuoltovarmuus-testissa/

Ofgem, 2023. All available charts. Office of Gas and Electricity Markets, London, UK. Accessed October 10, 2023: www.ofgem.gov.uk/energy-data-and-research/data-portal/all-available-charts

Onifade, S. T., Alola, A. A., Erdoğan, S., Acet, H., 2021. Environmental aspect of energy transition and urbanization in the OPEC member states. *Environmental Science and Pollution Research* 28, 17158–17169. https://doi.org/10.1007/s11356-020-12181-1

Osborne, S. P., 2006. The new public governance? *Public Management Review* 8, 377–387. https://doi.org/10.1080/14719030600853022

Overland, I., 2019. The geopolitics of renewable energy: Debunking four emerging myths. *Energy Research and Social Science* 49, 36–40. https://doi.org/10.1016/j.erss.2018.10.018

Pakalkaitė, V., Posaner, J., 2019. The Baltics: Between competition and cooperation, in: Godzimirski, J. (Ed.), *New Political Economy of Energy in Europe*. Springer, pp. 215–237. https://doi.org/10.1007/978-3-319-93360-3_9

Paltsev, S., 2016. The complicated geopolitics of renewable energy. *Bulletin of the Atomic Scientists* 72, 390–395. https://doi.org/10.1080/00963402.2016.1240476

Patel, S., 2023. Hydropower plant in Norway partly Collapses due to floods. *Power Magazine* 10.8.2023. Accessed September 11, 2023: www.powermag.com/hydropower-plant-in-norway-partly-collapses-due-to-floods/#:~:text=Record%2Dhigh%20river%20levels%20stemming,Eco%27s%20Braskereidfoss%20hydroelectric%20power%20plant

Pearson, P., Watson, J., 2012. UK Energy Policy 1980–2010: A history and lessons to be learnt. The Institution of Engineering and Technology, London, UK.

Peoples, C., Vaughan-Williams, N., 2015. *Critical Security Studies: An Introduction*. Routledge, Abingdon, Oxon.

Pesu, M., 2017. Koskiveneellä kohti valtavirtaa: Suomen puolustuspolitiikka kylmän sodan lopusta 2010-luvun kiristyneeseen turvallisuusympäristöön. Publications of the Ministry of Defence 1/2027, Helsinki, Finland. Accessed October 10, 2023: https://julkaisut.valtioneuvosto.fi/handle/10024/79901

Pesu, M., Iso-Markku, T., 2020. The deepening Finnish-Swedish security and defence relationship, FIIA Briefing Paper 291. Finnish Institute for International Affairs, Helsinki, Finland. Accessed October 10, 2023: www.fiia.fi/wp-content/uploads/2020/09/bp291_the_finnish-swedish_security_and_defence_relationship.pdf

Pflugmann, F., De Blasio, N., 2020. The geopolitics of renewable hydrogen in low-carbon energy markets. *Geopolitics, History, and International Relations* 12, 9–44. https://doi.org/10.22381/GHIR12120201

Piirimäe, K., 2020. From an "Army of Historians" to an "Army of Professionals": History and the strategic culture in Estonia. *Scandinavian Journal of Military Studies* 3, 100–113. https://doi.org/10.31374/SJMS.37

Piotrowski, C. S., 2018. Security policy of the Baltic states and its determining factors. *Security and Defence Quarterly* 22, 46–70. https://doi.org/10.5604/01.3001.0012.7586

PMO, 2009. Finnish security and defence policy. Prime Minister's Office, Finland. Accessed October 10, 2023: https://vnk.fi/julkaisu?pubid=3740

PMO, 2016. Government Report on Finnish Foreign and Security Policy. Prime Minister's Office Publications 9/2016, Finland.

Pöntinen, P., 2023. Itärajasta ei tule tuuliparatiisia. Suomen kuvalehti 23.1.2023. Accessed October 10, 2023: https://suomenkuvalehti.fi/kotimaa/takaisku-ita-suomelle-ministerioiden-selvitysmies-tuulipuistojen-tutkahaittaa-mahdotonta-ratkaista-teknisesti/

Prause, G., Tuisk, T., Olaniyi, E. O., 2019. Between sustainability, social cohesion and security regional development in northeastern Estonia. *Entrepreneurship and Sustainability Issues* 6, 1235–1254. https://doi.org/10.9770/jesi.2019.6.3(13)

Prosekov, A. Y., Ivanova, S. A., 2018. Food security: The challenge of the present. *Geoforum* 91, 73–77. https://doi.org/10.1016/J.GEOFORUM.2018.02.030

Qu, Y., Hooper, T., Swales, J. K., Papathanasopoulou, E., Austen, M. C., Yan, X., 2021. Energy-food nexus in the marine environment: A macroeconomic analysis on offshore wind energy and seafood production in Scotland. *Energy Policy* 149, 112027. https://doi.org/10.1016/j.enpol.2020.112027

Quitzow, R., Thielges, S., 2022. The German energy transition as soft power. *Review of International Political Economy* 29, 598–623. https://doi.org/10.1080/09692290.2020.1813190

Raik, K., Rikmann, E., 2021. Resisting domestic and external pressure towards de-Europeanization of foreign policy? The case of Estonia. *Journal of European Integration* 43, 603–618. https://doi.org/10.1080/07036337.2021.1927011

Raven, R., Van Den Bosch, S., Weterings, R., 2010. Transitions and strategic niche management: Towards a competence kit for practitioners. *International Journal of Technology Management* 51, 57–74. https://doi.org/10.1504/ IJTM.2010.033128

Regilme, S., Hartmann, H., 2019. Global shift, in: Romaniuk, S., Thapa, M., Marton, P. (Eds.), *The Palgrave Encyclopedia of Global Security Studies*. Palgrave Macmillan, Cham. https://doi.org/10.1007/978-3-319-74336-3_53-2

Resmonitor EU, 2021. Conflicts between national security reasons and wind energy development in Estonia. Resmonitor EU 25.10.2021. Accessed October 5, 2023: https:// resmonitor.eu/en/ee/barriers/286/

Righettini, M. S., Lizzi, R., 2022. How scholars break down "policy coherence": The impact of sustainable development global agendas on academic literature. *Environmental Policy and Governance* 32, 98–109. https://doi.org/10.1002/eet.1966

Rip, A., Kemp, R., 1998. Technological change, in: Rayner, S., Malone, E. L. (Eds.), *Human Choice and Climate Change. Vol. II, Resources and Technology*, Springer, Cham, pp. 327–399.

Ritchie, N., 2016. Nuclear identities and Scottish independence. *The Nonproliferation Review* 23, 653–675. https://doi.org/10.1080/10736700.2017.1345517

Rock, M., Murphy, J. T., Rasiah, R., van Seters, P., Managi, S., 2009. A hard slog, not a leap frog: Globalization and sustainability transitions in developing Asia. *Technological Forecasting and Social Change* 76, 241–254. https://doi.org/10.1016/j .techfore.2007.11.014

Roe, P., 2008. The "value" of positive security. *Review of International Studies* 34, 777– 794. https://doi.org/10.1017/S0260210508008279

Roe, P., 2012. Is securitization a "negative" concept? Revisiting the normative debate over normal versus extraordinary politics. *Security Dialogue* 43, 249–266. https://doi .org/10.1177/0967010612443723

Rogers-Hayden, T., Hatton, F., Lorenzoni, I., 2011. "Energy security" and "climate change": Constructing UK energy discursive realities. *Global Environmental Change* 21, 134–142. https://doi.org/10.1016/j.gloenvcha.2010.09.003

Rogge, K. S., Johnstone, P., 2017. Exploring the role of phase-out policies for low-carbon energy transitions: The case of the German Energiewende. *Energy Research and Social Science* 33, 128–137. https://doi.org/10.1016/j.erss.2017.10.004

Rogge, K. S., Reichardt, K., 2016. Policy mixes for sustainability transitions: An extended concept and framework for analysis. *Research Policy* 45, 1620–1635. https://doi .org/10.1016/J.RESPOL.2016.04.004

Rosenbloom, D., Rinscheid, A., 2020. Deliberate decline: An emerging frontier for the study and practice of decarbonization. *WIREs Climate Change* 11, e669. https://doi .org/10.1002/wcc.669

Runhaar, H., Wilk, B., Driessen, P., Dunphy, N., Persson, Å., Meadowcroft, J., Mullally, G., 2020. Policy integration, in: Biermann, F., Kim, R. (Eds.), *Architectures of Earth System Governance: Institutional Complexity and Structural Transformation*. Cambridge University Press, Cambridge, pp. 183–206.

Runhaar, H., Wilk, B., Persson, Å., Uittenbroek, C., Wamsler, C., 2018. Mainstreaming climate adaptation: Taking stock about "what works" from empirical research worldwide. *Regional Environmental Change* 18, 1201–1210. https://doi.org/10.1007/ s10113-017-1259-5

Ruostetsaari, I., 2010. Changing regulation and governance of Finnish energy policy making: New rules but old elites? *Review of Policy Research* 27, 273–297. https://doi .org/10.1111/j.1541-1338.2010.00442.x

Ruostetsaari, I., 2017. Stealth democracy, elitism, and citizenship in Finnish energy policy. *Energy Research & Social Science* 34, 93–103. https://doi.org/10.1016/j.erss.2017.06.022

Russel, D., Jordan, A., 2009. Joining up or pulling apart? The use of appraisal to coordinate policy making for sustainable development. *Environment and Planning A* 41, 1201–1216. https://doi.org/10.1068/a4142

Russel, D. J., den Uyl, R. M., de Vito, L., 2018. Understanding policy integration in the EU – Insights from a multi-level lens on climate adaptation and the EU's coastal and marine policy. *Environmental Science and Policy* 82, 44–51. https://doi.org/10.1016/j.envsci.2017.12.009

Safarzyńska, K., Frenken, K., Van Den Bergh, J. C. J. M., 2012. Evolutionary theorizing and modeling of sustainability transitions. *Research Policy* 41, 1011–1024. https://doi.org/10.1016/j.respol.2011.10.014

Santos Ayllón, L. M., Jenkins, K. E. H., 2023. Energy justice, just transitions and Scottish energy policy: A re-grounding of theory in policy practice. *Energy Research & Social Science* 96, 102922. https://doi.org/10.1016/j.erss.2022.102922

Scholten, D., 2018. The geopolitics of renewables – An introduction and expectations, in: Scholten, D. (Ed.), *The Geopolitics of Renewables*, Springer, pp. 1–33. https://doi.org/10.1007/978-3-319-67855-9_1

Scholten, D., Bazilian, M., Overland, I., Westphal, K., 2020. The geopolitics of renewables: New board, new game. *Energy Policy* 138, 111059. https://doi.org/10.1016/j.enpol.2019.111059

Scholten, D., Bosman, R., 2016. The geopolitics of renewables; exploring the political implications of renewable energy systems. *Technological Forecasting and Social Change* 103, 273–283. https://doi.org/10.1016/j.techfore.2015.10.014

Schönach, P., 2021. Rajattoman ilmakehän saastumista ja suojelua, in: *Suomen Ympäristöhistoria: 1700-Luvulta Nykypäivään*. Vastapaino, Tampere.

Schot, J., Geels, F. W., 2008. Strategic niche management and sustainable innovation journeys: Theory, findings, research agenda, and policy. *Technology Analysis and Strategic Management* 20, 537–554. https://doi.org/10.1080/09537320802292651

Schot, J., Kanger, L., 2018. Deep transitions: Emergence, acceleration, stabilization and directionality. *Research Policy* 47(6), 1045–1059. https://doi.org/10.1016/j.respol.2018.03.009

Schot, J., Steinmueller, W. E., 2018. Three frames for innovation policy: R&D, systems of innovation and transformative change. *Research Policy* 47, 1554–1567. https://doi.org/10.1016/j.respol.2018.08.011

Scott, W. R., 2001. *Institutions and Organizations. Ideas, Interests and Identities*. SAGE Publications.

Scottish Government, 2022. Energy statistics for Scotland – Q2 2022. Scottish Government, Edinburgh, UK. Accessed October 6, 2023: www.gov.scot/publications/energy-statistics-for-scotland-q2-2022/pages/renewable-electricity-capacity/

Scottish Government, 2023. Draft energy strategy and just transition plan – Delivering a fair and secure zero carbon energy system for Scotland. Scottish Government, Edinburgh, UK. Accessed October 10, 2023: www.gov.scot/publications/draft-energy-strategy-transition-plan/

Scottish Renewables, 2023. Statistics. Accessed October 6, 2023: www.scottishrenewables.com/our-industry/statistics

Sharples, J. D., 2016. The shifting geopolitics of Russia's natural gas exports and their impact on EU-Russia gas relations. *Geopolitics* 21, 880–912. https://doi.org/10.1080/14650045.2016.1148690

Shove, E., Walker, G., 2010. Governing transitions in the sustainability of everyday life. *Research Policy, Special Section on Innovation and Sustainability Transitions* 39, 471–476. https://doi.org/10.1016/j.respol.2010.01.019

Siddi, M., 2018. The role of power in EU–Russia energy relations: The interplay between markets and geopolitics. *Europe – Asia Studies* 70, 1552–1571. https://doi.org/10.1080/09668136.2018.1536925

Sillak, S., Kanger, L., 2020. Global pressures vs. local embeddedness: The de- and restabilization of the Estonian oil shale industry in response to climate change (1995–2016). *Environmental Innovation and Societal Transitions* 34, 96–115. https://doi.org/10.1016/j.eist.2019.12.003

Silvester, B. R., Fisker, J. K. 2023. A relational approach to the role of the state in societal transitions and transformations towards sustainability. *Environmental Innovation and Societal Transitions* 47, 100717. DOI: 10.1016/j.eist.2023.100717

Sivonen, M. H., Kivimaa, P., 2023. Politics in the energy-security nexus: An epistemic governance approach to the zero-carbon energy transition in Finland, Estonia, and Norway. *Environmental Sociology*, available online. https://doi.org/10.1080/23251042.2023.2251873

Skjølsvold, T. M., Ryghaug, M., Throndsen, W., 2020. European island imaginaries: Examining the actors, innovations, and renewable energy transitions of 8 islands. *Energy Research & Social Science* 65, 101491. https://doi.org/10.1016/j.erss.2020.101491

Smith, A., Raven, R., 2012. What is protective space? Reconsidering niches in transitions to sustainability. *Research Policy* 41, 1025–1036. https://doi.org/10.1016/j.respol.2011.12.012

Smith, A., Voß, J. P., Grin, J., 2010. Innovation studies and sustainability transitions: The allure of the multi-level perspective and its challenges. *Research Policy* 39, 435–448. https://doi.org/10.1016/J.RESPOL.2010.01.023

Smith Stegen, K., 2015. Heavy rare earths, permanent magnets, and renewable energies: An imminent crisis. *Energy Policy* 79, 1–8. https://doi.org/10.1016/j.enpol.2014.12.015

Sörlin, S., Dale, B., Keeling, A., & Larsen, J. N. 2022. Patterns of Arctic extractivism: Past and present, in: Sörlin, S. (Ed.), *Resource Extraction and Arctic Communities: The New Extractivist Paradigm*. Cambridge University Press, Cambridge and New York, pp. 35–65.

Sovacool, B. 2019. The precarious political economy of cobalt: Balancing prosperity, poverty, and brutality in artisanal and industrial mining in the Democratic Republic of the Congo. *The Extractive Industries and Society* 6(3), 915–939. https://doi.org/10.1016/J.EXIS.2019.05.018

Sovacool, B. K., Hook, A., Martiskainen, M., Baker, L., 2019. The whole systems energy injustice of four European low-carbon transitions. *Global Environmental Change* 58, 101958. https://doi.org/10.1016/j.gloenvcha.2019.101958

Sovacool, B. K., Mukherjee, I., 2011. Conceptualizing and measuring energy security: A synthesized approach. *Energy* 36, 5343–5355. https://doi.org/10.1016/J.ENERGY.2011.06.043

Sovacool, B. K., Turnheim, B., Martiskainen, M., Brown, D., Kivimaa, P., 2020. Guides or gatekeepers? Incumbent-oriented transition intermediaries in a low-carbon era. *Energy Research and Social Science* 66, 101490. https://doi.org/10.1016/j.erss.2020.101490

Statistics Estonia, 2021. Ida-Viru county. Accessed November 12, 2021: www.stat.ee/en/find-statistics/statistics-region/ida-viru-count

Statistics Estonia, 2023. Energy. Accessed October 4, 2023: www.stat.ee/en/find-statistics/statistics-theme/energy-and-transport/energy

Statistics Finland, 2022. Share of energy imported from Russia 34 per cent of total energy consumption in 2021. Accessed October 5, 2023: www.stat.fi/en/publication/cl1xmekvw1pp80buvn1cznxmy

Statistics Finland, 2023. Energian hankinta ja kulutus. Accessed October 5, 2023: www.stat.fi/tilasto/ehk

Statistics Norway, 2023a. Production and consumption of energy, energy balance and energy account. Accessed October 5, 2023: www.ssb.no/en/statbank/table/11561/

Statistics Norway, 2023b. Electricity. Accessed October 5, 2023: www.ssb.no/en/energi-og-industri/energi/statistikk/elektrisitet

Stenvall, J., Kinder, T., Kuoppakangas, P., Laitinen, I., 2018. Unlearning and public services – A case study with a Vygotskian approach. *Journal of Adult and Continuing Education* 24(2), 188–207. https://doi.org/10.1177/1477971418818570

Stirling, A., 2019. How deep is incumbency? A "configuring fields" approach to redistributing and reorienting power in socio-material change. *Energy Research & Social Science* 58, 101239. https://doi.org/10.1016/j.erss.2019.101239

Strambo, C., Nilsson, M., Månsson, A., 2015. Coherent or inconsistent? Assessing energy security and climate policy interaction within the European Union. *Energy Research and Social Science* 8, 1–12. https://doi.org/10.1016/j.erss.2015.04.004

Štreimikiene, D., Strielkowski, W., Bilan, Y., Mikalauskas, I., 2016. Energy dependency and sustainable regional development in the Baltic States – A review. *Geographica Pannonica* 20, 79–87. https://doi.org/10.5937/GeoPan1602079S

Studemeyer, C. C., 2019. Cooperative agendas and the power of the periphery: The US, Estonia, and NATO after the Ukraine crisis. *Geopolitics* 24, 787–810. https://doi.org/10.1080/14650045.2018.1496911

Szulecki, K. (Ed.), 2018a. *Energy Security in Europe: Divergent Perceptions and Policy Challenges*. Palgrave Macmillan, Cham. https://doi.org/10.1007/978-3-319-64964-1

Szulecki, K., 2018b. The multiple faces of energy security: An introduction, in: Szulecki, K. (Ed.), *Energy Security in Europe: Divergent Perceptions and Policy Challenges*. Palgrave Macmillan, Cham, pp. 1–30.

Szulecki, K., Kusznir, J., 2018. Energy security and energy transition: Securitisation in the electricity sector, in: Szulecki, K. (Ed.), *Energy Security in Europe Divergent Perceptions and Policy Challenges*. Palgrave Macmillan, Cham, pp. 107–148.

Szulecki, K., Westphal, K., 2018. Taking security seriously in EU energy governance: Crimean shock and the energy union, in: Szulecki, K. (Ed.), *Energy Security in Europe*. Palgrave Macmillan, Cham, pp. 177–202.

Tafon, R., Howarth, D., Griggs, S., 2019. The politics of Estonia's offshore wind energy programme: Discourse, power and marine spatial planning. *Environment and Planning C: Politics and Space* 37, 157–176. https://doi.org/10.1177/2399654418778037

Toke, D., Vezirgiannidou, S.-E., 2013. The relationship between climate change and energy security: Key issues and conclusions. *Environmental Politics* 22, 537–552. https://doi.org/10.1080/09644016.2013.806631

Tõnurist, P., 2015. Framework for analysing the role of state owned enterprises in innovation policy management: The case of energy technologies and Eesti Energia. *Technovation* 38, 1–14. https://doi.org/10.1016/J.TECHNOVATION.2014.08.001

Tosun, J., Lang, A., 2017. Policy integration: Mapping the different concepts. *Policy Studies* 38, 553–570. https://doi.org/10.1080/01442872.2017.1339239

Trombetta, M. J., 2009. Environmental security and climate change: Analysing the discourse. *Cambridge Review of International Affairs* 21, 585–602. https://doi.org/10.1080/09557570802452920

Truffer, B., Rohracher, H., Kivimaa, P., Raven, R., Alkemade, F., Carvalho, L., Feola, G., 2022. A perspective on the future of sustainability transitions research. *Environmental*

*Innovation and Societal Transitions* 42, 331–339. https://doi.org/10.1016/J.EIST.2022.01.006

Tuathail, G. Ó., 1999. Understanding critical geopolitics: Geopolitics and risk society. *The Journal of Strategic Studies* 22, 107–124.

Turnheim, B., Geels, F. W., 2012. Regime destabilisation as the flipside of energy transitions: Lessons from the history of the British coal industry (1913–1997). *Energy Policy, Special Section: Past and Prospective Energy Transitions – Insights from History* 50, 35–49. https://doi.org/10.1016/j.enpol.2012.04.060

Turnheim, B., Geels, F. W., 2013. The destabilisation of existing regimes: Confronting a multi-dimensional framework with a case study of the British coal industry (1913–1967). *Research Policy*, 42, 1749–1767. https://doi.org/10.1016/j.respol.2013.04.009

Turnheim, B., Kivimaa, P., Berkhout, F., 2018. *Innovating Climate Governance Moving beyond Experiments*. Cambridge University Press, Cambridge.

Tynkkynen, V.-P., 2021. *The Energy of Russia: Hydrocarbon Culture and Climate Change*. Edward Elgar Publishing, Cheltenham.

Umbach, F., 2010. Global energy security and the implications for the EU. *Energy Policy* 38, 1229–1240. https://doi.org/10.1016/j.enpol.2009.01.010

UN, 1987. *Report of the World Commission on Environment and Development: Our Common Future*. United Nations, Geneva.

UN, 1994. *Human Development Report 1994: New Dimensions of Human Security*. United Nations, Geneva.

Upham, P., Kivimaa, P., Mickwitz, P., Åstrand, K., 2014. Climate policy innovation: A sociotechnical transitions perspective. *Environmental Politics* 23. https://doi.org/10.1080/09644016.2014.923632

Vakulchuk, R., Øverland, I., Scholten, D., 2020. Renewable energy and geopolitics: A review. *Renewable and Sustainable Energy Reviews* 122, 109547. https://doi.org/10.1016/j.rser.2019.109547

Van de Graaf, T., 2018. Battling for a shrinking market: Oil producers, the renewables revolution, and the risk of stranded assets, in: Scholten, D. (Ed.), *The Geopolitics of Renewables*. Springer International Publishing, Cham, pp. 97–121. https://doi.org/10.1007/978-3-319-67855-9_4

Van de Graaf, T., Overland, I., Scholten, D., Westphal, K., 2020. The new oil? The geopolitics and international governance of hydrogen. *Energy Research & Social Science* 70, 101667. https://doi.org/10.1016/j.erss.2020.101667

van der Laak, W. W. M., Raven, R. P. J. M., Verbong, G. P. J., 2007. Strategic niche management for biofuels: Analysing past experiments for developing new biofuel policies. *Energy Policy* 35, 3213–3225. https://doi.org/10.1016/j.enpol.2006.11.009

van der Vleuten, E., 2019. Radical change and deep transitions: Lessons from Europe's infrastructure transition 1815–2015. *Environmental Innovation and Societal Transitions* 32, 22–32. https://doi.org/10.1016/J.EIST.2017.12.004

van Driel, H., Schot, J., 2005. Radical innovation as a multilevel process: Introducing floating grain elevators in the Port of Rotterdam. *Technology and Culture* 46(1), 51–76.

van Mierlo, B., Beers, P. J., 2020. Understanding and governing learning in sustainability transitions: A review. *Environmental Innovation and Societal Transitions* 34, 255–269. https://doi.org/10.1016/J.EIST.2018.08.002

van Oers, L., Feola, G., Runhaar, H., Moors, E., 2023. Unlearning in sustainability transitions: Insight from two Dutch community-supported agriculture farms. *Environmental Innovation and Societal Transitions* 46, 100693. https://doi.org/10.1016/j.eist.2023.100693

Vanttinen, P., 2022. Estonia develops its first offshore wind farms. *Euractiv* 7.11.2022. Accessed October 4, 2023: www.euractiv.com/section/politics/short_news/estonia-develops-its-first-offshore-wind-farms/

Varho, V., 2007. Calm or storm? Wind power actors' perceptions of Finnish wind power and its future, Environmentalica Fennica. Doctoral thesis, University of Helsinki, Helsinki. Accessed October 10, 2023: https://helda.helsinki.fi/items/9864c1bf-979a-483e-ab32-e1b81dba5041

Vasser, M., 2021. Estonia: From shale to gale. *Acid News* 16–17.

Verbong, G. P. J., Geels, F. W., 2010. Exploring sustainability transitions in the electricity sector with socio-technical pathways. *Technological Forecasting and Social Change* 77, 1214–1221. https://doi.org/10.1016/j.techfore.2010.04.008

Vesa, J., Gronow, A., Ylä-Anttila, T., 2020. The quiet opposition: How the pro-economy lobby influences climate policy. *Global Environmental Change* 63, 102117. https://doi.org/10.1016/j.gloenvcha.2020.102117

Victor, D. G., Jaffe, A. M., Hayes, M. H., 2006. *Natural Gas and Geopolitics: From 1970 to 2040*. Cambridge University Press, Cambridge.

Vihma, A., Reischl, G., Andersen, A. N., 2021. A climate backlash: Comparing populist parties' climate policies in Denmark, Finland, and Sweden. *The Journal of Environment and Development* 30, 219–239. https://doi.org/10.1177/10704965211027748

von der Leyen, U., 2023. Statement by President von der Leyen at the joint press conference with Norwegian Prime Minister Støre, NATO Secretary-General Stoltenberg and CEO Opedal, 17.3.2023. Accessed October 4, 2023: https://ec.europa.eu/commission/presscorner/detail/en/statement_23_1723

Wæver, O., 1995. Securitization and desecuritization, in: *On Security*. Columbia University Press, New York, pp. 46–87.

Wæver, O., 2011. Politics, security, theory. *Security Dialogue* 42(4–5), 465. https://doi.org/10.1177/0967010611418718

Willrich, M., 1976. International energy issues and options. *Annual Review of Energy* 1, 743–772.

Wilson, J. D., 2018. Whatever happened to the rare earths weapon? Critical materials and international security in Asia. *Asian Security* 14, 358–373. https://doi.org/10.1080/14799855.2017.1397977

Wilson, S., 2022. Energy firms dominate Insider's Top500 Scottish companies. *Insider* 25.2.2022. Accessed October 6, 2023: www.insider.co.uk/special-reports/energy-firms-dominate-insiders-top500-26298171

Winzer, C., 2012. Conceptualizing energy security. *Energy Policy* 46, 36–48. https://doi.org/10.1016/J.ENPOL.2012.02.067

Wither, J. K., 2020. Back to the future? Nordic total defence concepts. *Defence Studies* 20, 61–81. https://doi.org/10.1080/14702436.2020.1718498

Wolfers, A., 1952. "National Security" as an ambiguous symbol. *Political Science Quarterly* 67, 481–502. https://doi.org/10.2307/2145138

Wood, M., 2017. *Depoliticisation: What Is It and Why Does It Matter?* Sheffield Political Economy Research Institute, Sheffield.

Wrange, J., Bengtsson, R., 2019. Internal and external perceptions of small state security: The case of Estonia. *European Security* 28, 449–472. https://doi.org/10.1080/09662839.2019.1665517

Yates, J., 2022. *The National Grid Rip-Off*. Tribune, London.

Yazar, M., Haarstad, H., 2023. Populist far right discursive-institutional tactics in European regional decarbonization. *Political Geography* 105, 102936. https://doi.org/10.1016/j.polgeo.2023.102936

Yergin, D., 1988. Energy security in the 1990s. *Foreign Affairs* 67(1), 110–132.
Yergin, D., 2009. It's still the one – ProQuest. *Foreign Policy* 174, 89–95.
Ylönen, M., Litmanen, T., Kojo, M., Lindell, P., 2017. The (de)politicisation of nuclear power: The Finnish discussion after Fukushima. *Public Understanding of Science* 26, 260–274. https://doi.org/10.1177/0963662515613678
Żuk, P., Szulecki, K., 2020. Unpacking the right-populist threat to climate action: Poland's pro-governmental media on energy transition and climate change. *Energy Research and Social Science* 66, 101485. https://doi.org/10.1016/j.erss.2020.101485

# Index

accelerating transitions, 4, 21
actors, 7, 18–27, 29–31, 38, 42–44, 54–57, 60–66, 81–82, 92–93, 98, 100, 102–103, 107, 109–110, 115, 124, 131, 139, 147, 150, 158, 166, 176, 185
administration, 11, 61, 82–83, 101–103, 105, 109, 111, 113–114, 118, 145, 153, 156, 168
administrative, 7, 20, 59–60, 62, 73, 80, 82, 107, 114, 124, 130, 144, 165–166, 168
adverse sovereignty, 137, 139
Arctic, 57, 104, 106, 127, 129, 149, 151
armed conflict, 4, 31, 42, 49

Baltic, 2, 8–9, 46, 49–50, 69–70, 77, 79, 81, 84, 90–91
Baltic connector, 70, 174
Belize, 149
bioenergy, 94, 97, 114
biofuel, 23, 75, 97
Booth, Ken, 6, 29–30, 139
border, 8–9, 40–41, 69, 76, 85–86, 92, 95, 104, 106, 113, 117, 122, 127, 135, 151
Britain, 146, 149–150

Canada, 149
capacity, 1–2, 30–31, 44, 49, 81, 88, 90, 97–98, 102, 123, 125, 130, 132, 143, 152, 154
carbon dioxide, 133, 141, 166
Cherp, Aleh, 7, 25, 36–38
China, 2, 5, 42, 45, 87, 106, 114, 128, 155, 158, 172
citizen energy ownership, prosumer, 1
"Climate and Energy Strategy," 101, 106, 111, 115–116
climate change, 8, 13, 25–28, 34, 36–37, 40, 47, 50–51, 54, 55, 57, 60, 92, 98–100, 103–104, 115, 117, 128, 131, 139, 149, 151, 156, 159, 161, 166, 185
Climate Change Act, 125, 141, 144–145, 147–148
climate diplomacy, 185
climate policy, 11, 47, 74–75, 85–86, 96, 144, 148
coal, 2, 9, 34, 70, 96, 99–100, 106, 133, 141–142, 144, 175

coalition, 18, 96, 148
co-evolution, 20, 176
comprehensive security, 25, 102–103
conflict, 2, 7, 25, 32, 42, 48, 59, 73, 80–81, 88–89, 114, 166–167, 172
consensus, 65, 79, 87–88, 90, 93, 98, 152, 157, 167, 176, 181, 183
Copenhagen School, 26–27, 32, 169–170
Crimea, 49, 77, 86, 96, 109–110, 129, 181
crisis, 26, 38, 45–47, 70, 83–84, 87, 90–91, 96, 104–105, 108–109, 114–115, 121, 136, 144, 147, 152
critical infrastructure, 84, 135, 137, 171
critical materials, 2, 41, 43, 45, 85, 92, 112, 114, 118, 132, 155, 161–162, 168, 172, 179
critical minerals, 84, 119, 155, 160, 174, 179
"Critical Minerals Strategy," 149, 155, 167
cultural, 18–20, 22–33, 37, 43, 47, 54, 56, 62, 65, 78–79, 95, 100, 102, 106, 108, 134, 151, 182
cybersecurity, 2, 11, 77–78, 81, 92, 108, 127, 130, 150, 171, 173

dealignment, 20
decarbonization, 2, 4, 7, 36–37, 40–41, 46, 50–51, 76, 80, 82, 85, 87, 93, 97, 111–112, 117, 132, 141, 144–145, 154, 178
defence forces, 77
defence policy, 8, 13, 53, 60, 62, 77–78, 81, 102, 106–107, 111, 117, 127, 141, 149–151, 153, 160, 165–166, 185
defence radar, 9, 118, 166
deinstitutionalization, 66, 134
democracy, 4, 20, 27, 43
Denmark, 2, 38, 49
depoliticization, 28–29, 32, 118, 146, 154, 160, 169, 184
desecuritization, 9, 27–28, 102, 117, 169
destabilization, 24, 33, 52, 63–64, 72, 79, 84–85, 92, 111–112, 132, 138, 158, 175
desynchronization, 46, 72–73, 81, 84, 90–91, 167, 172

211

digital technologies, 2, 45, 56, 70
diplomacy, 47, 153, 160
discourse, 24, 27, 110, 111, 130, 137–138, 140
disruption, 20, 24, 26, 31, 36–37, 39, 41–43, 47, 49–50, 53, 55, 64, 70, 78, 91, 95, 104, 107, 115, 129, 134, 136, 151
dissonance, 105, 107, 121, 139
Dupont, Claire, 28–29

ecological, 20, 76
economic security, 9, 13, 36, 87, 121, 123, 125, 132–133, 166, 175, 178
economy, 181
Eesti Energia, 73–75, 85, 87–88, 178
electricity
  distribution, 1, 18, 41, 107, 126, 146
  transmission, 1, 10, 18, 46, 73–74, 81, 91, 107, 120, 123, 125, 130, 146, 178
electrification, 1–3, 9, 13, 42, 44, 98, 112, 118, 123, 125, 172–173, 178, 186
Elering, 73–74, 82, 91, 169
emissions, 36–37, 75, 82, 85, 87, 97, 115, 121, 125–126, 133, 136, 141, 166, 175, 185
empowerment, 21–23, 46
energy crisis, 84, 122, 170, 174, 177
energy democracy, 171
energy dependence, 1, 50, 110
energy diplomacy, 185
energy efficiency, 1, 9, 59, 73, 76, 99, 109, 138, 141, 144–147, 153–154, 156, 186
energy elite, 74, 98, 104
energy independence, 8, 13, 38, 76, 85, 92, 110, 130, 172–173, 175
energy justice, 33, 147, 174
energy policy, 3, 7, 9–12, 29, 32, 35, 38, 40, 47, 49, 51, 57–59, 62, 66, 69, 73–74, 76, 79–81, 87, 92, 94, 96, 99, 101–103, 105–112, 117–118, 121–122, 124–127, 130–132, 137–138, 140–141, 144–148, 151–155, 157, 159–162, 165–166
energy poverty, 38, 147, 154, 160, 162, 173
energy saving, 17, 51, 111
energy security, 7, 11–13, 29, 32, 35–39, 41, 43, 48–50, 75, 83, 85, 92, 100, 108, 111–112, 114–116, 121, 123, 127, 129–130, 133–134, 138, 140, 142, 144, 147, 152–158, 161–162, 167, 174, 186
energy–security nexus, 176, 178–179, 183
energy sovereignty, 137, 140, 179
energy transitions, 3–13, 18, 31, 34, 40, 42, 46, 47, 51, 54, 56, 59, 63, 72, 96, 108, 112, 114, 118, 130, 139, 143, 157, 161, 165, 168, 170, 171, 172, 173, 177, 178, 184
environment, 6, 19, 21, 23, 25–26, 28–29, 36–38, 47, 54–56, 58, 72, 77, 87, 98, 100–101, 106–107, 133, 135, 148, 153–154, 159
environmental security, 25–26, 38–39, 54
Equinor, 123–126, 131, 138, 178

Estonia, 2–3, 8–11, 13, 40, 66, 69–82, 84–94, 112, 127–129, 131, 135, 138, 165–166, 169, 171–173, 182
Estonian, 172, 176, 178, 181, 183. *See* Estonia
EU, 3, 8, 28, 43, 49–51, 60, 69, 71, 74, 76–77, 80–81, 85, 87, 90, 104, 106, 110, 122, 124–125, 138, 144, 172
EU Green Deal, 58, 185
Europe, 1–2, 8–9, 35, 45–48, 54, 61, 70, 79–80, 90, 92, 106, 110, 116–117, 121, 123, 125, 131, 133, 135–138, 142, 152, 167, 185
European, 1, 3, 5–6, 8–9, 21, 29, 33, 38, 49–50, 55, 60–61, 69–70, 75, 81, 83, 87, 90, 94, 96, 104, 108, 117, 119, 121, 123–124, 129, 133–134, 136, 138, 144–146, 149, 155, 158, 168
European Commission, 6, 51, 60, 83, 90, 181
European Union. *See* EU
expectations, 21, 22, 30, 32–33, 63, 65, 114, 128, 162
experiment, 21–23, 41
experimentation, 19, 23–24, 118

far-right, 86, 88, 173
Fennovoima, 110–111
Finland, 2, 3, 8–11, 13, 35, 40, 66, 69–70, 72, 79, 81, 82, 90–91, 93–100, 102–106, 108–119, 122, 128–130, 135, 138, 159, 165–166, 169
Finlandization, 95, 106, 108, 112, 117, 182
Finnish, 10, 25, 49, 70, 73, 96–108, 112, 115–118, 169, 176, 178, 181, 183, 185
fit-and-conform, 23, 175
Floyd, Rita, 6, 25–28, 30
foreign affairs, 10, 73–74, 76, 81, 102, 110, 127–128, 130, 156–157
foreign policy, 47, 49, 105, 127, 149–150, 152, 154, 156, 172
fossil fuels, 1, 9, 32, 41, 44–45, 59, 75, 93, 96, 98, 103, 105, 114, 116, 121, 123, 133, 141, 147, 154, 156, 168

gas, 9, 36–38, 40, 48–50, 65, 70, 72–73, 75, 82–84, 90, 93, 95–97, 99, 106, 111–112, 115–116, 118–126, 128–129, 131–133, 135–136, 138–139, 141–143, 145–147, 149, 151–152, 154, 160, 166, 171
Gazprom, 70
Geels, Frank, 5, 18–22, 24, 32, 52, 54–55, 64–65
geopolitical/geopolitics, 2–3, 5, 7–8, 11–12, 32–36, 38–48, 51, 54, 56–58, 69, 77, 84, 89, 92–93, 98, 103, 105, 106, 108, 110–111, 114, 116–118, 128, 131–132, 134, 139, 152–153, 155–156, 162, 166, 172, 176–177, 183–184
Germany, 2, 5, 47, 49–50, 112, 116–117, 122
Global North, 172, 184
Global South, 172, 184
governance, 6, 23, 30–31, 58, 60, 74, 77, 86, 98, 101, 109, 146
grid communities, 2, 46, 92

hard security, 77, 173, 184
health, 6, 26, 37–38, 107, 162
heating, 1, 9, 71, 75, 97, 100, 107, 115–116, 141, 154
Hoogensen Gjørv, Gunhild, 4, 6, 26–27, 29–30, 33, 77, 161, 170
horizontal coherence, 60, 153
Hungary, 49
hydrocarbon, 2, 7–9, 40–48, 98, 120–122, 132, 138, 160–161, 172, 175
hydrogen, 1, 7, 40, 51, 75, 112, 123, 125–126, 143
hydropower, 9, 38, 42–44, 66, 94, 97, 119–125, 129–135

Ida-Viru County, 79–81, 85–88, 90, 92, 175
incumbent, 21, 23–24, 32, 64, 66, 98, 101, 108
Indigenous, 18, 47
industrial policy, 126
inequality, 25
infrastructure, 8, 33, 35–36, 38, 42, 54, 81, 107, 117, 130, 133, 136, 154, 159
injustice, 24, 33
innovation, 2, 5, 19, 20, 22–23, 33–34, 64–65, 94, 101, 118, 144, 155, 162
innovation policy, 94, 101
insecurity, 30–31, 86, 139, 143, 170–171
institutional change, 48, 57, 81, 93
institutionalization, 21, 23, 29
institutionalize, 21, 23, 127
institutions, 4, 19, 42, 47–48, 58, 126, 151
International Energy Agency (IEA), 2, 48, 119, 151, 172
international relations, 6, 25, 41
Iraq, 149
IRENA, 3, 48

Jewell, Jessica, 7, 25, 36, 37, 38
Johnstone, Phil, 3, 5, 17, 24, 31, 32, 33, 35, 40, 47, 144, 145, 153, 159, 169
justice, 2, 4
Just Transition Commission, 143, 148, 155, 171, 175
Just Transition Mechanism, 87, 171, 175
just transitions, 25, 34, 87, 92, 116, 143, 148, 155, 161, 162, 171, 173, 184

Kemp, René, 5, 19, 20, 55
Kenya, 149

landscape, 19–22, 32, 37, 47, 52–57, 62, 79–80, 98, 103–107, 109, 117, 129, 143–144, 151, 165, 181
learning, 8, 21–23, 61, 64–65, 162, 166, 172, 175
liberalization, 48, 50, 144
LNG, 70, 82–84, 93, 96, 111, 117, 167
lock-in, 58, 154

military, 2–3, 5–6, 8, 25, 29, 31–33, 35–37, 39–40, 54–56, 58, 77–79, 89, 104, 105, 112, 127–129, 131, 133–135, 137–138, 140, 143, 149–150, 153, 155–156, 159–160, 166, 168, 173, 176, 184–185
military security, 2–3, 25, 31
Ministry of Defence, 76, 86, 88–90, 92, 102, 127, 149–150, 156, 177
MLP, 19, 20–21, 52–53

National Grid, 146–147, 158, 161
national security, 7–9, 13, 25–27, 31, 34, 60, 62, 69, 77, 81, 84–85, 88–90, 92, 109, 112, 114, 127, 131, 149, 153, 160, 173
National Security Council, 157–158
NATO, 8, 49, 69, 77, 85–86, 108, 127, 129, 149, 151, 153, 166, 173
negative security, 6, 29–33, 52–53, 63, 69, 72, 77, 92, 96, 114, 116–117, 122, 135, 137–139, 143, 158, 161, 165, 170–171
network, 8–10, 18, 21–22, 43, 64, 73, 78, 90–92, 102, 107–108, 116, 120, 123, 126, 134, 141, 146, 156, 172
niche development, 22, 24, 32, 66, 72, 79, 83–84, 96, 111–112, 122, 131, 143, 158, 162
niche expansion, 23, 84, 92–93, 112, 120, 125, 138, 162, 165, 173, 175, 182–183
niche innovations, 5, 20–23, 32, 52, 58, 172
non-military, 25–26, 77, 149
Nord Pool, 2, 46, 90, 138
Nord Stream, 38, 50–51, 70, 111–112, 116, 118, 135, 137–138, 152, 174
Norsk Hydro, 123, 134, 173
Norway, 10–11, 13, 40, 44, 46, 66, 94, 97, 119–140, 143, 165–166, 169
Norwegian, 9, 65, 119–128, 130–138, 140, 142, 171, 178
nuclear power, 3, 9, 72, 94, 97, 99, 110–112, 117, 129, 147, 152–153, 155, 158–160, 162, 169, 176

OECD, 6, 60, 75
offshore wind, 65, 72, 73, 81, 88, 97, 113, 125, 131, 140, 158, 160, 175
Ofgem, 145, 178
oil, 8–9, 13, 32–33, 35–36, 40–41, 44, 48–49, 55, 57, 65–66, 69–76, 79–88, 90–93, 96–97, 99, 106, 114, 119–126, 128–133, 135–136, 138–139, 141–143, 149, 153, 160
oil shale, 8, 13, 66, 69–76, 79–88, 90–93, 166, 167, 171–172, 178
Olkiluoto 3, 3, 97
OPEC, 48, 119
organizational, 8, 18, 61, 127, 144, 157

parliament, 9–11, 49, 127, 145, 149
parliamentary, 58, 61, 77, 109–110, 145, 149
peace, 4, 6–7, 30, 33–34

peacebuilding. *See* peace
peat, 9, 66, 94, 96, 100, 111, 112, 114–117, 166, 171, 175
phaseout, 8, 24, 64, 66, 72, 84, 86–87, 92, 95, 115, 122–123, 141, 157, 160, 166, 171–172, 175, 182
Poland, 2, 5, 49–50, 90–91
policy coherence, 5, 7, 12, 21, 49, 51–53, 58–62, 72, 81, 96, 109, 122, 129, 143, 157, 160, 165, 168, 185
policy domains, 7, 13, 59–62, 80–81, 107, 149, 166, 168, 170
policy incoherence, 7, 60, 62, 87, 107, 117, 168, 183
policy instruments, 107, 153, 162
policy integration, 5, 7–8, 58, 60–61, 107, 109, 155, 160, 165–166, 168
policy measures, 29, 62, 80
policy mix(es), 20, 58–59
policy objective, 7, 59–62, 80, 89, 168
policy processes, 59, 80, 157
political, 3, 6, 17–20, 26–29, 31, 34–37, 40, 46, 49, 54, 56–59, 61–62, 65–66, 74, 79, 82–85, 87–88, 90, 92–93, 97–98, 100–102, 105, 108–110, 112, 115, 117, 125, 131–133, 137–138, 140, 146–148, 150–152, 154, 157, 160, 162
political incoherence, 183
politicization, 28, 140, 165, 169–170
politicize, 28, 83, 124, 130, 137
politics, 6–7, 10, 19, 27–28, 32, 34, 40, 44, 48, 77, 86, 93, 105–106, 171
pollution, 17, 34, 36, 37
populism, 34, 46, 54
populist, 34, 46, 77, 86, 171
positive security, 4, 6, 27, 30–33, 46, 92, 102, 114, 116–118, 122, 132, 136–137, 139–140, 161–162, 165, 170–172, 175
power, 18, 29, 31, 39–41, 44, 46, 64, 91, 98, 145–146
Power Pool, 100, 107, 166, 179
practices, 17, 19, 23, 28–31, 47, 58, 62, 65, 75, 109, 140

rare earths, 2
Raven, Rob, 21–23
reconfiguration, 17, 21, 24, 84, 112, 182
referent object, 25–26, 55, 116, 183
regime decline, 23, 63–64, 66, 134, 139, 165, 172, 175, 182–183
regulation, 37, 88, 135, 148
renewable energy, 1–4, 7, 9, 12–13, 32, 35, 38, 40–49, 56, 59, 60, 71–76, 79, 83–84, 88, 91–93, 96, 100, 108, 112, 120, 122–124, 126, 134, 137, 141, 143–144, 146–147, 152, 154–155, 158, 168, 171, 173–174, 177–179, 181, 186
replication, 23
RePower EU, 51, 83

resilience, 31, 37, 102, 171, 174, 184–185
reskilling, 143, 158
resource colonialism, 172
restabilization, 182
Rip, Arie, 5, 19–20, 55
Roe, Paul, 4, 6, 29–31
Romania, 49
Rosatom, 3, 105, 110
rules, 7, 19, 23–24, 33, 46–47, 56, 74
Russia, 1, 3, 5, 8–9, 11, 21, 33, 40, 46, 48–51, 54–55, 57, 69–70, 72, 76–82, 84–88, 90–96, 98–99, 102–119, 121, 122, 127–129, 131, 133, 135, 137, 143–144, 149–153, 155, 157, 161–162, 166, 170

safety, 6, 26, 30, 97, 107, 122, 134, 151, 155, 158
Schot, Johan, 18–22, 32, 34, 55
Scotland, 2, 3, 8–11, 13, 66, 141–143, 145–148, 150–152, 157–158, 160, 162, 165–166, 171, 175
Scottish, 9, 143, 145–148, 151–153, 155, 157–158, 159–161
securitization, 4, 26–29, 32, 53, 62, 96, 111, 122, 138, 140, 143, 165, 169–170
security, climate, 2, 62, 103, 132, 139, 154, 156, 161
security, human, 2, 25–26, 30, 38, 54, 172
security of supply, 2, 13, 43, 45, 48, 50, 54–56, 58, 73, 81, 87–88, 90, 100, 102, 108–110, 115–116, 118, 120, 129, 136, 146, 154, 158, 165, 167, 174, 177, 179
security policy, 11, 13, 29, 32, 45, 47, 50, 59, 60, 62, 76–77, 80, 93, 102–103, 105, 107, 116, 127, 130–131, 135, 149–150, 155, 165
security studies, 3, 5–7, 12, 18, 25–29, 32, 36–37, 55, 138, 169–170, 183–184
social justice, 6, 18, 30, 173
sociotechnical, 5, 17–21, 23–24, 32–33, 52, 54, 56, 175, 182
soft security, 184
solar, 23, 42–44, 51, 66, 71, 75, 83, 92, 131, 134, 147
Sovereign Wealth Fund, 171, 175
Soviet Union, 69, 102, 105, 110, 169, 182
stability, 13, 30–31, 34, 40–41, 52, 56, 81, 106, 112, 133, 138, 153
state, 8, 18, 25–26, 30–35, 38, 40–41, 46, 55, 58, 69, 73–75, 77, 79, 82, 106, 110, 114, 115, 122–124, 127, 129, 132, 136, 145, 149–150, 152
stockpiling, 36, 84, 112, 167–168, 171
stretch-and-transform, 23, 175
sustainability transitions, 3–6, 12–13, 17–21, 24–26, 31, 33–34, 37, 41, 47, 52, 54–56, 58, 63, 82, 118
Sweden, 2, 38, 90–91, 94, 97, 112, 119, 122
synergy, 7, 59–60, 62, 80, 83, 87, 107, 109, 131, 167
Szulecki, Kacper, 3, 7, 27, 29, 34, 37–38, 46, 49, 53, 62, 131

## Index

technological change, 19, 22, 42
technology, 6, 22–23, 33–34, 37, 41–43, 63, 81, 89, 125–126, 143, 149
tensions, 24, 33, 42, 44–46, 62, 78, 86, 135
territorial, 6, 70, 96, 102, 116, 122
terrorist, 3, 36, 39, 127, 130, 133, 158, 166, 174, 184
threats, 6, 13, 25–30, 32, 36, 58, 77, 78, 92, 104, 106
total security, 107
transformation, 20–21, 33, 44
transition pathways, 13, 20
transport, 1, 9, 17, 23, 32, 72, 97, 99, 123, 129, 151, 159

UK, 2, 8–9, 72, 122, 141, 144
Ukraine, 1, 3, 21, 33, 40, 48–50, 54–55, 70, 77, 80, 90, 93–95, 104–106, 110, 114–115, 121, 126, 135, 137, 144, 151–152, 155, 170, 181
UN, 26, 127, 129
US, 5, 34, 51, 72, 85, 111, 117, 119, 149
unlearning, 64–65, 168
upscaling, 23

Wæever, Ole, 27
wellbeing, 6, 25, 30, 38, 58, 86, 92, 107
Willrich, Mason, 36
wind power, 2, 8–9, 22, 33, 38, 44, 66, 71–72, 80–82, 84, 88–90, 92–93, 96–97, 111–114, 117–118, 120, 123, 125, 131, 134–140, 143, 147, 158, 160, 162, 166, 173, 175
world wars, 5, 32–35, 40
World War II, 32, 34, 96, 182

zero-carbon, 2–4, 8, 11–12, 54, 56, 62, 69, 81, 84–85, 88–89, 91–93, 111, 115, 121, 124, 130, 145–146, 160, 166–167, 175, 178, 185

Milton Keynes UK
Ingram Content Group UK Ltd.
UKHW012104151124
451073UK00027B/298